Animals & Men

THE PEAK DISTRICT WALLABIES
Animal Poltergeists; The Orang Pendek;
Tasmania Expedition Preliminary Report;
News. reviews and more...

The Journal of the Centre for Fortean Zoology   #56

Contents

Typeset by Jonathan Downes,
Cover and Layout by SPiderKaT for CFZ Communications
Using Microsoft Word 2000, Microsoft Publisher 2000, Adobe Photoshop CS.
First published in Great Britain by CFZ Press

CFZ Press, Myrtle Cottage, Woolsery, Bideford, North Devon, EX39 5QR

© CFZ MMXV

978-1-909488-44-1

Faculty of the Centre for Fortean Zoology

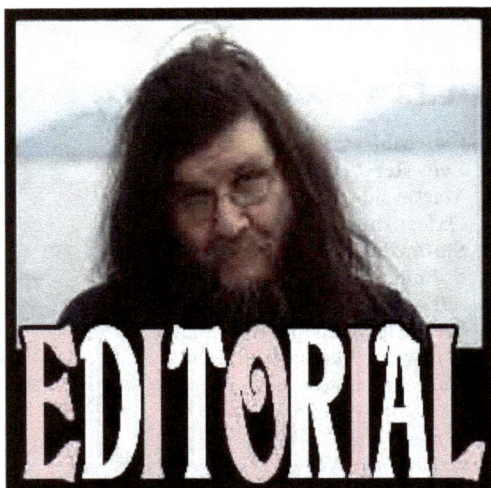

EDITORIAL

Dear Friends,

Welcome to the first edition of *Animals & Men* for 2016. I am still, as you see, trying to adhere to my promise of the 2014 Annual Report, that we would return to a publication schedule of four issues a year plus twelve issues of the monthly newsletter. This has been hampered slightly by losing my secretary. Jessica decided in January that she wanted to follow a different career path and left.

However the lovely Marianne comes in one afternoon a week to help me with my correspondence, and the CFZ continues on the same wobbly trajectory that it has done for the past quarter of a century. Yep, next year will see twenty five years since I sat on the shores of Loch Ness smoking a suspiciously long cigarette and decided that starting a club for people interested in mystery animals was not specifically that bad an idea.

Things have changed quite a lot in the past few years. Mainly, as I see it, as a result of the Government's vile austerity measures, people who once had spare time now have nothing of the sort.

Until a matter of days ago it looked as if for the first time in seventeen years there would be no Weird Weekend in Devon this year. There have been problems with the venue, and I have lost nearly all of my helpers. Ten years ago there was a team of twenty people, whereas now it is basically down to three or four. And finally my own health is declining at an alarming rate, so I had to make the very difficult decision as to whether we continue or not. However, we all think that it would be a great shame if the event, which so many people enjoy, fell victim to the apathy, prejudice and selfishness of a handful of people. So we are continuing - for this year at least.

However, we are very short-handed and would like to take this opportunity to appeal for volunteers to help in return for free entry. The Weird Weekend is not about money, and never has been, but it is nice if we can run it without being out of pocket. This year will see two events, with Glen Vaudrey's Weird Weekend North just after Easter and there are even plans for a Weird Weekend in Scandinavia next year.

So far the following speakers have been confirmed:

- Lars Thomas: The Vikings and their Monsters
- Steve Rider: tba
- Matt Cook: High Strangeness and hill forts
- Mick Walters: Werewolves in

The Great Days of Zoology are not done!

4

Staffordshire
- Julian Vayne: Chaos Magick
- Music from Stargrace
- Richard Freeman: Tasmania 2016 Expedition Report
- Joe Thomas: Cryptozoology on film
- Richard Muirhead: Devo and the Monkeyman
- Matthew Watkins: Retrocausality and other reverse-time phenomena
- Shoshannah McCarthy: tba
- Ronan Coghlan: tba
- Jon Downes: Keynote Speech

I have always been disappointed that wherever we have held the event; Exeter, Woolsery and now Hartland, very few local people have come along, and this is something that I have always wanted to rectify. If anyone reading this has any sensible suggestions for how we can broaden our audience I would be very grateful to receive them.

In fact, much the same thing could be said about the CFZ. Whereas we are still, as far

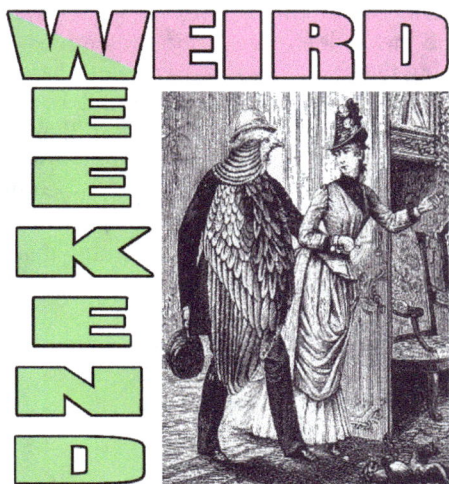

WEIRD WEEKEND

Une Semaine de Freaky

WEIRD WEEKEND

Une Semaine de Freaky

as I am aware, the largest mystery animal research group in the world, we have lost members in recent years due to the internecine infighting of the Internet, and to the advent of other groups and organisations that come and go, and offer things ideologically that we shy away from. The fact that cryptozoology has become of interest to the mass media in recent years due to the prominence of various reality TV shows broadly on the subject has brought an influx of new people into the scene, many of whom find our workmanlike attitude to be impossibly staid.

But the existence of this issue of *Animals & Men*, which features writing from the new generation of cryptozoologists like Carl Marshall and Liam Dorricott, proves that there are young people for whom the discipline is not about rushing around woodlands in combat fatigues talking nonsense about 'squatches with cloaking

19-21 August 2016
The Small School, Hartland,
Devon
www.weirdweekend.org

devices. And this is something for which I feel truly grateful.

I have written at length, in these pages and elsewhere, about how concerned I am becoming that whole generations of our society are becoming more and more divorced from the reality of the natural world, but I never once considered that due largely to the machinations of the crappier end of the television industry, that in many people's eyes the discipline would be akin to a crudely authored video game.

Back when I took my career as a musician and composer seriously, I was continually angry at the fact that the venues that I and my management would approach with a view to my band performing there, would continually prefer to book cover bands, karaoke or even strippers rather than the sort of mildly cerebral art rock that we played and sang. Why were people so unutterably vulgar and stupid? I would moan.

However, in recent years I have found myself thinking much the same thing about the Fortean disciplines within which I have been working for such a long time. In the same way as my band never did even an ironic cover version of _Agadoo_ in order to provide a sop to the _hoi polloi_, I am going to continue to ignore requests to

write profiles of the stars of the latest reality TV show claiming to be hunting for Bigfoot, and I truly cannot see a world in which I shall be changing my mind.

Please forgive me for having spent the content of another editorial bellyaching, but it has been a long, hard year so far and I cannot truthfully see it getting any better any time soon.

I was talking to Jaki Windmill at last year's Weird Weekend, and we agreed that we are living in times where we don't know anyone who is actually having a good time. It is as if we are rapidly ending a period of darkness for us all. If we are, I hope that I don't sound too pretentious in saying that it us up to us all to work harder than ever to spread the light. And I truly am not meaning to sound either like an Old Testament prophet or a self-serving TV evangelist. Merely like a desperately worried cripple in late middle age who is confronted by a whole lot of things that I neither like, accept or understand.

Thank you to everyone who has stood by me, and continues to stand by me as we continue on our journey.

Jon Downes

A LEGAL MATTER

Wherever possible we use images that are either owned by us, public domain, or with the permission of the copyright holder. However, when we are unable to do this we believe that we are justified under the Fair Use legislation.

Copyright Law fact sheet P-09 : Understanding Fair Use
http://www.copyrightservice.co.uk/copyright/p09_fair_use
Issued: 5th July 2004

What is fair use?

In copyright law, there is a concept of fair use, also known as; free use, fair dealing, or fair practice. Fair use sets out certain actions that may be carried out, but would not normally be regarded as an infringement of the work. The idea behind this is that if copyright laws are too restrictive, it may stifle free speech, news reporting, or result in disproportionate penalties for inconsequential or accidental inclusion.

What does fair use allow?

Under fair use rules, it may be possible to use quotations or excerpts, where the work has been made available to the public, (i.e. published). Provided that: The use is deemed acceptable under the terms of fair dealing. That the quoted material is justified, and no more than is necessary is included. That the source of the quoted material is mentioned, along with the name of the author.

Typical free uses of work include:

Inclusion for the purpose of news reporting, incidental inclusion.

National laws typically allow limited private and educational use.

This magazine is not produced for profit, it is free to read and share. The hard copy version is available through Amazon and other outlets, but on a not for profit basis. We feel that we are justified in our use of copyrighted materials under several of the above clauses.

Newsfile

Monitoring the situation

Separated by several hundred kilometres from its next of kin, a new species of blue-tailed monitor lizard unique to the remote Mussau Island has been described. Unknown to science until recently and formally termed the "isolated," it is the only large-sized land-living predator and scavenger native to the island. Dubbed a "biogeographical oddity" by its discoverers, led by Valter Weijola, a graduate student from the University of Turku, Finland, the lizard species is also the first new monitor lizard to be described from the country of Papua New Guinea in over twenty years. The finding was published in the open-access journal *ZooKeys*.

SOURCE: http://tinyurl.com/zjl73qn

Turning Turtle

A new species of freshwater turtle has been discovered in Papua New Guinea, a mountainous and tropical country in the south Pacific Ocean.

The turtle is one of three distantly-related species found across Papua New Guinea and neighbouring parts of Indonesia. A study published in the international journal, *Zootaxa*, says the turtle, named *Elseya Rhodini*, is one of three distantly-related species found across PNG and Indonesia's Papua province.

The study's lead author, Arthur Georges from the University of Canberra, says the three species evolved from a common ancestor somewhere between 17 and 19 million years ago. "That animal has seen the whole orogenesis [mountain creation] of New Guinea happen," he said.

SOURCE: http://tinyurl.com/jj6u5jb

The Lion sleeps Tonight

A previously unknown population of at least 100 lions has been discovered by a wildlife charity in a remote park in north-western Ethiopia. Born Free Foundation said it had obtained camera trap images and identified lion tracks in the Alatash area close to the border with Sudan. The area is thought to have lost all its lions in the 20th Century because of hunting and habitat destruction. The number of lions in Africa has declined by half since the 1990s.

The lions are thought to be of the Central African sub-species, of which only about 900 were thought to survive, Born Free Foundation's programmes manager, Mark Jones, told the BBC Newsday programme. "Even though the team only visited the Ethiopian side of the park because of logistics, lions were likely to exist in the larger, adjacent Dinder National Park across the border in Sudan," he said.

SOURCE: http://tinyurl.com/h2ojtk6

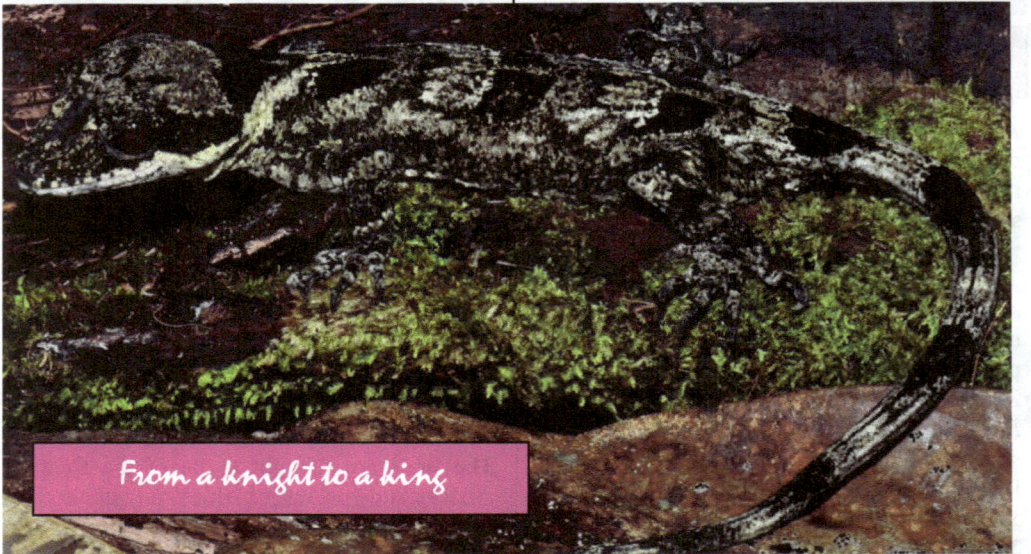

From a knight to a king

The extremely complex geological history of New Guinea has allowed many of its animals and plants the chance to grow different enough to make a name for themselves. Two new gecko species whose names respectively mean 'knight' (equestris - above) and 'king' (rex - below) were discovered by a team led by Dr. Paul Oliver, The Australian National University and University of Melbourne, are described in the open-access journal ZooKeys. Both new species belong to the world's most diverse gecko genus Cyrtodactylus which comprises more than 200 species known to date.

SOURCE: http://tinyurl.com/jm6rfos

A tree frog thought to be extinct for more than a century has been rediscovered in the jungles of north-east India. The frog genus Frankixalus has been rediscovered "in abundance" by a group of scientists led by renowned Indian biologist Sathyabhama Das Biju, who is known as The Frog Man in India. However, the team believe they remain under threat with the region's tropical forests disappearing because of programs to cut trees, plant rice, expand human settlements and build roads, AP reports. "This is an exciting find, but it doesn't mean the frogs are safe," Biju said. "This frog is facing extreme stress in these areas, and could be pushed to extinction simply from habitat loss," Biju added. "We're lucky in a way to have found it before that happens, but we're all worried."

SOURCE: http://tinyurl.com/hcn79re

Two endemic tree frog species, not recognised by science until now, have been identified in broadleaf forests in the island country of Taiwan. Unlike their siblings from mainland China and Southern Asia, they demonstrate reproductive behaviour, characterised with egg-eating (oophagous) tadpole embryos feeding on eggs, while still inside the mother's womb. What told them apart initially, however, were their gemstone-coloured eyes. The research team, led by Dr. Shu-Ping Wu, University of Taipei, have their study published in the open-access journal *ZooKeys*.

SOURCE http://tinyurl.com/hld4k9p

Found on a herb bush, a toad of only 24 mm average length, measured from its snout tip to its cloaca, was quick to make its discoverers consider its status as a new species. After identifying its unique morphological and skeletal characters, and conducting a molecular phylogenetic analysis, not only did Dr. Aggarwal, Centre for Cellular & Molecular Biology, Dr. Vaudevan, Wildlife Institute of India and Laboratory for Conservation of Endangered Species along with their team, introduce a new species, but also added a new genus. The new 'Andaman bush toad', is described in a paper published in the open-access journal *ZooKeys*.

SOURCE: http://tinyurl.com/jkc2csg

Another Compendium of Batrachia

A team of scientists including a Smithsonian Tropical Research Institute (STRI) research associate announced the discovery of a new species of pale-gold coloured frog from the cloud forests of the high Andes in Colombia. Its name, *Pristimantis dorado*, commemorates both its color (dorado means 'golden' in Spanish) and El Dorado, a mythical city of gold eagerly sought for centuries by Spanish conquistadores in South America.

"The Spaniards assumed Colombia's wealth was its gold, but today we understand that the real riches of the country lie in its biodiversity," said Andrew Crawford, a STRI research associate and faculty member at the Universidad de Los Andes.

The extraordinarily diverse group to which the new species belongs, Pristimantis, includes 465 recognized species, 205 of them from Colombia. The mountainous terrain of the Andes probably led to the evolution of so many different ground-dwelling frogs, in which the eggs develop directly into tiny baby frogs without going through a tadpole phase. At seven-tenths of an inch long, the species is among the smaller species in the group. The largest species grow to be 2 inches in length.

SOURCE: http://tinyurl.com/houyhmu

A team of researchers from India and the National University of Singapore (NUS) has discovered a new species of narrow-mouthed frog in the laterite rock formations of India's coastal plains. The frog, which is the size of a thumbnail, was named *Microhyla laterite* after its natural habitat. The discovery by the research team, led by Mr Seshadri K S, a PhD student from the Department of Biological Sciences at the NUS Faculty of Science, was published in the journal *PLOS One* on 9 March 2016. Laterite rock formations are prominent landscape features in the coastal plains of southwest India.

They are broadly considered as rocky areas as they are usually devoid of trees and other vegetation, and are therefore classified as wastelands. While conducting field surveys as a part of his citizen science initiative "My laterite, My habitat," independent researcher Mr Ramit Singal, who is one of the authors of the journal paper, spotted the frog in laterite habitats in and around the coastal town of Manipal, Udupi District, Karnataka State, India. He brought it to the attention of Mr Seshadri and his collaborators, who worked together to describe the frog.

SOURCE: http://tinyurl.com/j4kyanx

Another Compendium of Batrachia

Pokemon octopod

In February, during its first operational dive of 2016, the NOAA's Okeanos Explorer discovered a new octopus species that has been described as everything from an adorable cartoon character to a ghostly, Pokémon-like cephalopod, agency officials announced earlier this week. The Okeanos Explorer craft was exploring the waters off northeastern of Necker Island (Mokumanamana) in the Hawaiian Archipelago on February 27 as part of a deep dive to help determine whether or not there was a connection between the island and Necker Ridge, a narrow feature which extends more than 400 miles and protrudes beyond the current US exclusive economic zone.

In addition, its remotely operated vehicle (ROV) Deep Discoverer had been surveying biological communities at depths of nearly 4,300 meters, when it located the mysterious octopod resting on a flat rock lightly covered in sediment, the NOAA said. Its appearance was unlike any previously known cephalopod, and its discovery was the deepest that such a creature had ever been seen.

SOURCE: http://tinyurl.com/zlcjd8z

Scientists get rattled

There are more species of rattlesnake slithering around western North America than previously thought according to a study by University of Arkansas biologists Michael Douglas and Marlis Douglas and their colleagues at the University of Illinois at Urbana-Champaign and Western Kentucky University. The researchers published their findings in the journal PLOS One. The research team recommend that six groups of subspecies of the western rattlesnake be elevated to full species status, with the following names:

Crotalus viridis, prairie rattlesnake
Crotalus oreganus, northern Pacific rattlesnake
Crotalus cerberus, Arizona black rattlesnake
Crotalus helleri, southern Pacific rattlesnake
Crotalus concolor, midget faded rattlesnake
Crotalus lutosus, great basin rattlesnake

The names will be submitted to the International Committee on Zoological Nomenclature for ratification.

SOURCE: http://tinyurl.com/gnvyprz

Arctic Role

Some might say it takes a rare breed to survive the Alaska wilderness. The discovery of a possible new species of hybrid butterfly from the state's interior is proving that theory correct. Belonging to a group known as the Arctics, the Tanana Arctic, *Oeneis tanana*, is the first new butterfly species described from the Last Frontier in 28 years and may be its only endemic butterfly. University of Florida lepidopterist Andrew Warren suggests the butterfly could be the result of a rare and unlikely hybridization between two related species, both specially adapted for the harsh arctic climate, perhaps before the last ice age. Details of the finding are available online in the Journal of Research on the Lepidoptera. Digging deeper into the Tanana Arctic's origins may reveal secrets about the geological history of arctic North America and the evolution of hybrid species, said Warren, who led the new study.

"Hybrid species demonstrate that animals evolved in a way that people haven't really thought about much before, although the phenomenon is fairly well studied in plants," said Warren, senior collections manager at the McGuire Center for Lepidoptera and Biodiversity at the Florida Museum of Natural History on the UF campus. "Scientists who study plants and fish have suggested that unglaciated parts of ancient Alaska known as Beringia, including the strip of land that once connected Asia and what's now Alaska, served as a refuge where plants and animals waited out the last ice age and then moved eastward or southward from there. This is potentially a supporting piece of evidence for that."

SOURCE: http://tinyurl.com/hxxoeyl

Elton's shrimp

The shrimp-like creature was found in Indonesia, and the scientist decided to name him after Sir Elton John. The crustacean, *L. eltoni,* cannot sing 'Rocket Man', so why did his discoverer name him after the famous singer? James Thomas from the Halmos College of Natural Sciences and Oceanography in Florida, who discovered the shrimp-like creature, explained "when this unusual crustacean with a greatly enlarged appendage appeared under my microscope after a day of collecting, an image of the shoes Elton John wore as the 'Pinball Wizard' came to mind."

SOURCE: http://tinyurl.com/hcquuus

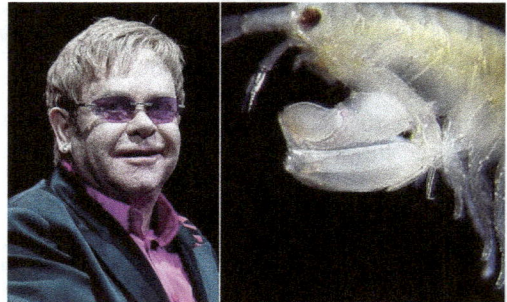

Chupacabras

Goat 2 Hell

After some months of the only chupacabras reports being nothing of the kind, on February 16th it appears that there was a return to the sort of vampiric attacks that were far more common back at the end of the 1990s, and which Graham Inglis and I investigated in Puebla during early 1998.

Farmer Xóchitl Zamorano, who owns the 13 sheep which were killed, alleges that he went to her yard upon hearing "strange noises" and a creature "shot out and hid in the shelter in the morning". Apparently all the animals were facing in the same direction as if they had been "intentionally aligned". However horses, cows and goats who were also there were untouched. The attacks took place in the village of La Palma, in the municipality of Angostura However until we receive any further information it is difficult to draw any conclusions from the scant (and badly translated) accounts and the unsubstantiated pictures.

If there are any Spanish speakers reading this who would like to look into this case further please contact me. Watch this space

SOURCES: http://tinyurl.com/hq2j79t
http://tinyurl.com/j5gnqrp

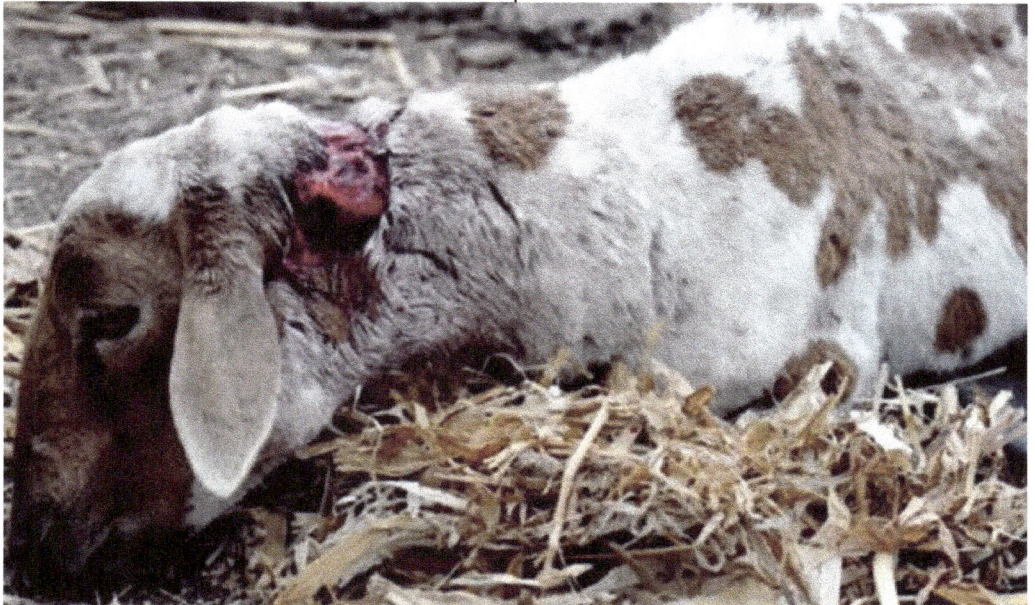

Man Beasts (BHM)

Sasquatch Chronicles

This interesting photograph comes from the Sasquatch Chronicles blog. The statement accompanying it reads: "As with any and all photo's I make no claims. This photo was sent to me and was taken from a person who works in Fish and Game. I am sharing it with you guys, sometimes it is better to get several eyes to look at a photo instead of just mine. I am trying to get more information. I would like to speak to the person on-air if I can get them to come on. I am working on it.

What do you guys and gals think? I viewed the picture on a large HD screen and the one thing that really caught my attention was the eye shine. I wanted to share it with you guys, take a look and let me know what you think."

SOURCE: http://tinyurl.com/zkgrhsf

Colorado Cryptid

The following still is taken from some game camera footage allegedly taken in Colorado

I have not been able to find out any more information whatsoever about the footage but look forward to receiving any more details if any sleuth amongst the A&M readership can help in the matter.

SOURCE: http://tinyurl.com/zswxz49

Pulling Mussels from the shell

The following story is another one from the increasingly impressive Sasquatch Chronicles blog: "I was out for a walk by myself on a nice spring day enjoying nature and getting some peace and quiet away from my children. I like to walk down to the Ninnescah River, so I went to the closest bridge by our house. I walk quietly so I can hear everything around me, I practice listening to everything and using all my senses when I walk. I got to the edge of the bridge at the river and I stopped as I saw a figure squatting down at the edge of the river.

First I thought it was a homeless person or local, but then I noticed they were covered in hair, so then I wondered why someone would be dressed in a bigfoot costume at this remote area by the river, then I looked around to see if there was anyone else filming trying to pull a hoax or something- you never know what the teenagers will do these days. There was no one, and that's when I started noticing other details that pointed to the reality of the situation. The bigfoot was breaking open mussels or clams or whatever and then eating them, which made me think 'Who would eat that raw from this dirty river where there is trash laying around?'. That's when it fully dawned on me that this was not a human. It was covered in a brown hair/fur. It was squatting with its back towards me the whole time, it never looked back at me, and I didn't move until I turned and left to go home to try to get my phone to take a picture (which it was dark by the time I got home and there was nothing there when I got back)."

SOURCE: http://tinyurl.com/jtgfcnr

Wisconsin Waderer

Another very useful web resource for those of us interested in BHM reports from around the world is the Bigfoot Evidence website. Recently it took a critical look at a suspiciously impressive slice of video footage.

They write:

"In this video, Phil Poling of Parabreakdown fame takes a look at an amazingly clear piece of bigfoot footage from the great state of Michigan. In this video a bigfoot is apparently cornered by a person who happens to have a video camera handy. Unlike many bigfoot videos, the subject is close, and fairly clear for part of the video. This unfortunately is also its downfall. Have a look: "

At the CFZ we are very impressed with the work that the Bigfoot Evidence and the Sasquatch Chronicles crews are doing. In a genre which is becoming increasingly muddied by hoaxing and well-meaning gullibility, these blogs are amongst the forerunners doing it the way it SHOULD be done.

SOURCE: http://tinyurl.com/gtndl25

Mystery Cats

UK Round Up

Big cat sightings in the UK have been few and far between this year so far. However there have been a few, although far fewer than usual:

YORKSHIRE
These pictures were taken at Low Marishes near

Malton. It appears to show a big cat in the field. They were taken by Highways gritter driver Michael Armitage. He's sent the pictures to North Yorkshire Police too and says he now believes there is something the rumours of Big Cats on the prowl because he's seen it for himself now.

SOURCE: http://tinyurl.com/z3twuk9

LANCASHIRE
NEW sightings of the 'Beast of Bolton' have been reported by terrified locals in the northern town. Several reports of a giant cat prowling the streets at night have been recorded by townsfolk, with the latest claiming the animal is the "size of a car bonnet".

Frightened eyewitness Natalie Kay came in contact with the huge creature while driving her Vauxhall Zafira at around 9.30pm.

"I had to slam the brakes on and it just stopped dead right in the middle of the road, staring at me," she recalled. "It was big with a long tail, pointed ears and greedy yellow eyes - about the same size as the bonnet of my car and really scary. You hear of these animals in jungles, but you wouldn't think anything like that could live in Bolton." She said the 'Beast' ran down a dirt track which leads to Higher and Lower Doe Hey Reservoirs, the ideal hiding spot.

SOURCE: http://tinyurl.com/j9bja9x

SUFFOLK
Eliot Evans, a sixth-former at Thomas Mills High School, in Framlingham, says he came face-to-face with a frighteningly large feline while out for an evening jog near his home in Wickham Market. "I had started to do some sprints and was coming down towards the tennis courts when I saw what I first thought was a large dog – but I also realised I was the only person on the field, so it couldn't be with its owner. It was still there, staring at me. I thought maybe my eyes were messing with me but I started to feel kind of anxious and sprinted across the pitch. I turned back to see it was moving towards me. At this point, I was growing nervous, so I shone the torch on my phone and saw two large eyes, too far apart for it to be any dog. It went prone and

began coming towards me again. It must have been between four and five feet long, at least." Eliot, who is studying biology, as well as maths, psychology and chemistry, has since carried out his own research on local big cat sightings.

SOURCE: http://tinyurl.com/h28a4ok

CFZ Suffolk representative Matt Salusbury followed up this sighting, and his account of his investigations can be found here:

http://tinyurl.com/zc3v724

CAMBRIDGESHIRE
John Walker, 71, claims to have spotted the so-called Fen Tiger hunting on the outskirts of Cambridge while he was walking his dog. The retired electronics engineer claims he had an unexpected brush with the fearsome creature - which was prowling the fields merely 100 yards away. He said: "At first I thought it might be a fox, but when my collie-cross gave chase, I realised that it had a very long and slender tail, which drooped downward then curled up in a wild cat-like shape. "The beast was moving in a lowered cat-like manner until it spotted my dog approaching, then it sprinted for cover, and Ginny heeded my call and returned to me. This cat was the same size as my dog, which weighs about 20kg (three stone) and it had a longer tail, and large paws. The coat was dark brown with lighter browns underneath, and it seemed to have a much darker muzzle. The field where the animal was has lots of rabbit holes in it, so maybe the beast, whatever it was, was hunting."

SOURCE: http://tinyurl.com/zac8b5d

PEEBLES
A big cat has again been reported prowling just yards from houses in St Andrews. Drew Lumsden spotted the beast as he drove daughter Kaitlyn, 7, from their home on the south-east edge of the town to her swimming class. The landscaper insisted the black animal was the size of an Alsatian or Labrador dog but had a feline gait and long tail. He said: "I am 100% certain it was a cat. It was certainly no dog." Big cat sightings have been reported in that area before, with a flurry in 2012. The animal spotted last

Thursday evening was in a field about 300 yards from the road near the Grange Inn. Drew said: "I was taking my daughter to swimming just before 6pm and she saw it first. She thought it was a fox at first but it was no fox, it was black. I could tell it was a cat by its head and the length of its tail. It was walking along, as casual as anything. I stopped to take a photograph but it was too dark and pictures came out grey."

SOURCE: http://tinyurl.com/zowxq8y

KENT
A large animal, thought to be a 'black Leopard', has been spotted near several Kent towns. Three sightings of the animal, or animals, have been reported on social media within 90 minutes of one another. One person, who lives in Lower Hardres near Canterbury, saw the creature just 20 metres from their back garden after being alerted by their pet Weimeraner dog. It was described as feline, larger than the agitated Weimeraner, and longer than a deer. Reports of a similar creature have also been posted by people living near Sevenoaks and Maidstone.

SOURCE: http://tinyurl.com/z7lrbej

CUMBRIA
Midwife Claire Greensmith spotted a "big, black panther-like creature" while hiking at a Lake District landmark. Now it's feared a black panther, or family of panthers, are ripping the heads of sheep throughout the famous English beauty spot. Cumbria Police have reassured holidaymakers and locals that sightings of 'The Beast' are rare and always investigated by officers. Mrs Greensmith, 41, explained how she was enjoying the stunning waterfall views at Colwith Force near Ambleside when she saw the creature by the nearby River Brathay. She said: "It was completely unexpected. It wasn't something I was going looking for but I saw what looked like a big, black cat with a long tail. It was definitely too big to be a normal cat and it wasn't a dog. I'd never heard about any of the other sightings until I got home and Googled it."

SOURCE: http://tinyurl.com/hoc9o7g

Aquatic Monsters

Back to Barmouth

I have to admit that I am never too sure about the *Daily Express* despite a mate of mine being a part of their senior editorial staff. However, the paper has brought out some interesting stories recently. There have been reports of a large marine creature seen in the vicinity of Barmouth on the west coast of Wales, for many years. However, they have died down in the last three decades.

The *Daily Express* reported on the 14th of January that:

"Llanarth pensioner Mohammad Tahla snapped the 'monster-like shape' in the estuary of the River Aeron, 60 miles away from Barmouth, sparking speculation that the Barmouth Monster has migrated south." "Mr Tahla said: "I took one look and thought 'Blimey, that looks just like the Loch Ness Monster',' he told the Cambrian News. "I went down there two days later to see if I could spot it again but by then it had gone." Of course, it is most likely that this was just a piece of driftwood or other detritus, but, it does have to be said that it looks remarkably like the reports of the Barmouth monster that has been reported over the years.

However, not all 'monsters; reported in the region have this squat, giant turtle like appearance. This 19th Century report written up by our friend Tabitca is testament to this:

"About three P.M. on Sunday, September 3, 1882, a party of gentlemen and ladies were standing at the northern extremity of Llandudno pier, looking

towards the open sea, when an unusual object was observed in the water near to the Little Orme's Head, travelling rapidly westwards towards the Great Orme. It appeared to be just outside the mouth of the bay, and would therefore be about a mile distant from the observers. It was watched for about two minutes, and in that interval it traversed about half the width of the bay, and then suddenly disappeared. The bay is two miles wide, and therefore the object, whatever it was, must have travelled at the rate of thirty miles an hour. It is estimated to have been fully as long as a large steamer, say two hundred feet; the rapidity of its motion was particularly remarked as being greater than that of any ordinary vessel. The colour appeared to be black, and the motion either corkscrew-like or snake-like, with vertical undulations. Three of the observers have since made sketches from memory, quite independently, of the impression left on their minds, and on comparing these sketches, which slightly varied, they have agreed to sanction the accompanying outline as representing as nearly as possible the object which they saw. The party consisted of W. Barfoot, J.P., of Leicester, F. J. Marlow, solicitor, of Manchester, Mrs. Marlow, and several others. They discard the theories of birds or porpoises as not accounting for this particular phenomenon.
F. T. MOTT.

Birstall Hill, Leicester, January 16th, 1883."

Mexican Globster

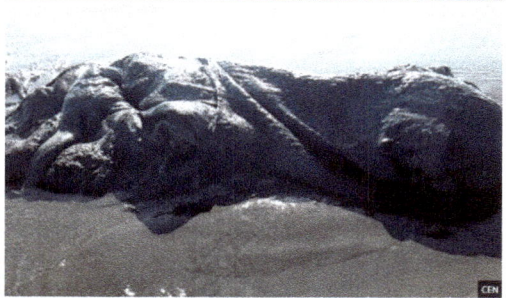

This particularly noisome object was washed up on a beach in Mexico on early March, and the story was sent to us by Indiana rep Elizabeth Clem, who has been a useful source for peculiar stories for years. *Maxim* (of all publications) wrote: "A mysterious four-meter long (13ft) beast was washed up on a beach in Mexico on Thursday, stunning beach-goers. They found the decomposing body on Bonfil Beach, in the city of Acapulco, in the south-west Mexican state of Guerrero. The news, as Mirror reports, immediately went viral on social media, and the sea-creature's photos were shared thousands of times."

Although the coordinator of the Civil Guard and Fire Brigade, Rosa Camacho, believes the animal had not been dead for a long time, it still seemed to have rapidly started to decay. She said: "We have no idea what type of animal this is, but I do know that it does not smell bad or have a fetid aroma. It is four metres long and was found on Bonfil Beach."

SOURCES: http://tinyurl.com/ja9xmd4, http://

Eel meat again

As I have said before in these pages, the way that people across the developed world seem to have become more and more divorced from the reality of the natural world, is something that increasingly disturbs me. I am, afraid, only too aware that I am becoming a boring old git, and partially so on this subject, but it is something which does matter immensely to me. In mid-February, for example, a photograph of a mysterious creature which had been washed up on a beach in New South Wales, was splashed across the internet.

The *Daily Telegraph* described the "monster from the deep" as like a giant eel with the head of the Frankenstein love-child of a crocodile and dolphin.

A "messed-up crocodile," a mystery lake monster and giant eel were some of the other descriptions given the sea monster. Some even suggested it was Photoshopped. The photo was taken by Robert Tyndall near the Swansea boat ramp on what is Australia's largest coastal saltwater lagoon in New South Wales. The 13-mile-long lake is connected to the Tasman Sea by a short channel. The sighting was said to be near the mouth of the lake. The hint here is in the last paragraph. Despite being called a lake, Lake Macquarie is actually a salt water lagoon. And various marine biologists have identified the creature as being a pike eel (*Muraenesox bagio*). I have to say that the creature in the photograph has a dome to it's skull not found on pike eels, but I am perfectly prepared to accept that as purely an optical illusion caused by the photographic angle. The way that the brain interprets images like this never ceases to amaze me, and if I had my time over again I think I would probably do a lot more studying of the psychology of perception.

SOURCE: http://tinyurl.com/jggyc49

Teratology

Thai mutant?

The calf with the head of a crocodile which was born in a remote village in Thailand last autumn is, of course, nothing of the sort. It is the body of a calf suffering from a well known if unusual skin disease. However, initial reports had villagers hailing it as some kind of divine omen. The animal was born in the village of High Rock by a buffalo that has previously given birth to normal litters. According to the site, it died soon after birth, and "is believed to bring good luck to the family and the village." Nhênyêt jiyanê uploaded photos to her page with one reader writing "praise to god."

SOURCE: http://tinyurl.com/jcu2jed

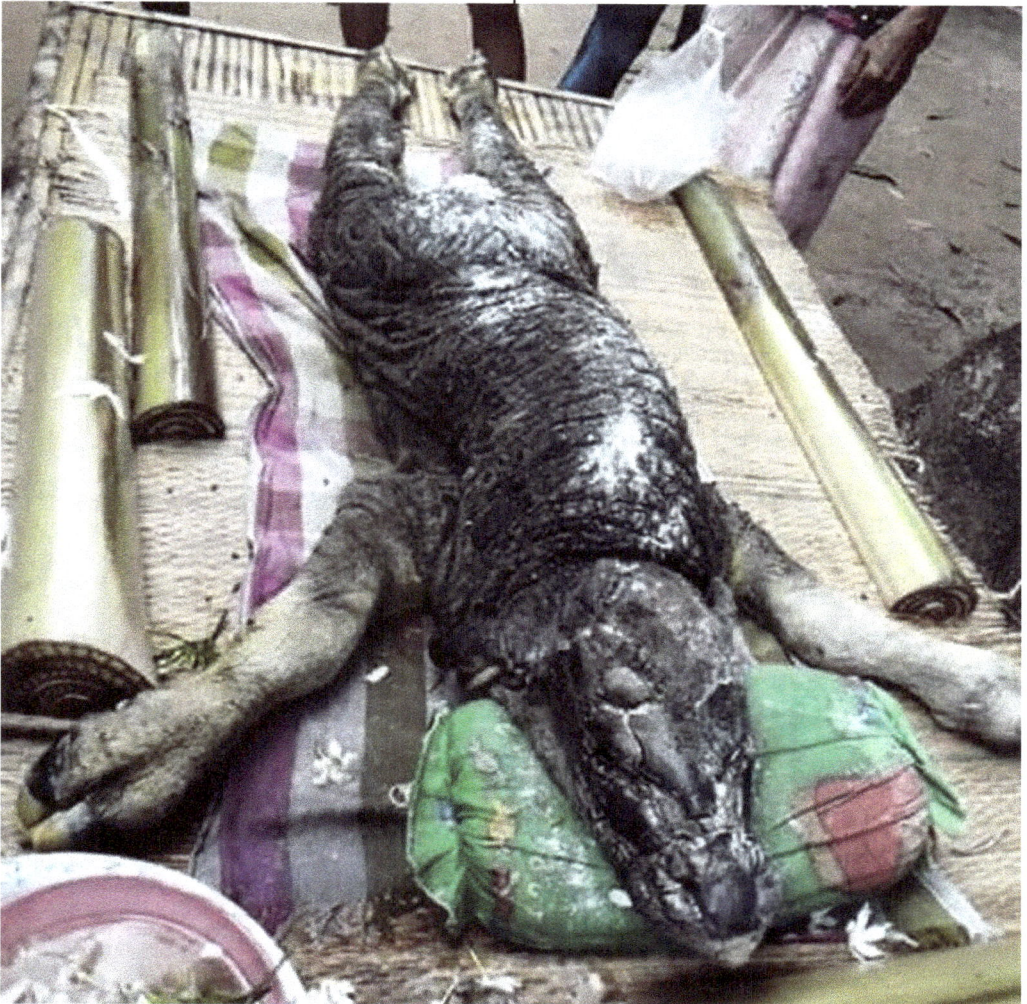

Odds and Sods

Werewolves of London again

There is an old saying about life imitating art. It was one of the things my mother used to say, but as she went on to the next plane of existence thirteen years ago now, I can't really ask her where it came from. Whereas when I first started the CFZ back in 1992 werewolf reports, like sighting of Springheeled Jack was something firmly consigned to the history books, in recent years there has been somewhat of a resurgence of them, particularly in the USA. However there have been few in Britain, especially in a, and we actually have a speaker booked for net years Weird Weekend. The most famous British werewolf accounts of modern times, are of course those associated with the Hexham Heads mystery with the Northumberland town of the same name. If you want to know more about that, by the way, I refer you to the book on the subject which is published by one of the imprints of CFZ publishing.

Back in the day when my first wife Alison and I were engaged on a surrealchemical quest masterminded by my old friend Tony Shiels, he insisted on playing the most famous song by the late Warren Zevon. And he played it over, and over, and over again. But I never thought that there actually would be a report of a werewolf from Britain's capital. Check out this story that appeared on This Is London at Yule. http://tinyurl.com/p73entw

The article begins:

Mystery surrounds a "haunting animal-like howling" which plagues sleeping residents of Belvedere. Sleepy folk claim the "scary moaning" can be heard across Clydesdale Way, Norman Road and Upper Belvedere every night and into the early morning. Chloe Philpot, 25, said: "Over the past month or so I have been hearing an extremely haunting animal-like howling."It is repetitive but each sound is never the same. "I know a few people in the area have tried to search for the source to no avail."It's getting scary."

The story concludes with miss Philpot likening the 'Pitiful' sound to that of a 'Mutant owl' as the story then goes on to name check the late Jim Morrison, I tend to think that it was probably a joke. However, you do never know. Although I think it's highly unlikely that I think it's a werewolf, I would urge all readers of this newsletter who live, or who have friends/relatives who live in the belvedere area of London to keep their ears peeled.

In January I wrote that "I highly doubt that it is a werewolf, but something interesting may be going on". Actually, according to Cryptomundo on March 20th nothing interesting was happening at all. Craig Woolheater writes: "Rowena Osmond, of Ripley Road, has revealed the real reason behind the disturbing noises. The 65-year-old said: "A woman went and stood underneath the wind turbines in Eastern Way, near to the Crossness sewage works, and heard the same noise. "They were very loud and the whole of Belvedere could hear them. But, I never believed that the noises were coming from a wolf for one second."

Chloe Philpot previously told *News Shopper* that the alleged howling was "getting scary" just before Christmas. The wind turbines are reportedly set to be serviced on soon where the loud noises will become a tale of the past."

Banal but true.

What is it with wallabies?

A pair of fishermen have found a dead wallaby in the River Exe in Devon. Matt Welham and Richard Williams caught the animal while trying to catch pike. The animal is believed to have been in the water for several days. It's thought to have escaped from a nearby farm.

Matt said: "As it was pike we were bottom fishing and I got a what seemed like a bite. I hauled it in and it looked at first as if there was a dead cat on the line. When it got closer and out of the water it looked like a kangaroo. I guessed it was stuffed but when we saw it close up we realised it was real and a wallaby. It must have weighed about 25lb and we reckoned it had been in the water for about two days tops."

The pair, who are both from Exeter, were fishing close to the Mill on the Exe pub in the city at the time.

SOURCE: http://tinyurl.com/jtx8cds

Jersey Devil? Surely not...

Dave Black of Little Egg Harbor Township was driving home from his security guard job in Atlantic City when he saw what he thought was a llama running in and out of the trees lining the road. "I was just driving past the golf course in Galloway on Route 9 and had to shake my head a few times when I thought I saw a llama," he wrote in his email.

What happened next is the bizarre part. "If that wasn't enough, then it spread out

leathery wings and flew off over the golf course." Black said he grabbed his cell phone and snapped off a few photos, but only one came out. The creature quickly disappeared, he said, and left Black wondering. "Either my mind is playing tricks on me or I just saw the Jersey Devil," he wrote.

After his encounter, Black said he stared at the photo for an hour trying to come up with an explanation for the image he had captured, before deciding to share it with NJ.com and its readers.

"Thought I'd send it in for you to share," he wrote. "I'm not looking for anything in return, just thought someone else could maybe explain this in a more rational way."

SOURCE: http://tinyurl.com/jfhkd9g

Waiting for the worms

Massive worms which grow to the size of a small snake and are as heavy as mice have been discovered on a "barren island" in Scotland.

Some of the animals were 16ins long and weighed a whopping 12.5g, making them the largest earthworms in Britain, according to researchers.

They were found in Papadil, an abandoned settlement on the Isle of Rum, and were three to four times bigger than the average worm. The project's lead researcher Dr Kevin Butt agreed that the findings were "slightly spooky", whilst speaking on BBC Radio 4's Today programme. However there's little to fear from the giant creatures, as earthworms tend to avoid people and will burrow into the ground if they hear footsteps. Scientists believe the colony's remote location and lack of predators such as moles, badgers, hedgehogs and foxes allowed them to grow to their impressive size.

The biggest specimen previously spotted weighed just 8g, whilst the average worm is around 4g-5g. The research was published in The Glasgow Naturalist, and conducted by a team of researchers from the University of Central Lancashire.

SOURCE: http://tinyurl.com/hrtsfcd

As you may have noticed, there are considerably more news items this issue than usual. As I have commented elsewhere, although having access to the world's media at the click of a mouse is a fascinating resource, it is also a mixed blessing. When I went to do the news pages for this issue I found so many things that I wanted to put in that I couldn't fit them into the normal space allocated in the magazine for news stories. So I made the decision to try and catch up with ourselves, which is why there are more news stories in this issue, and less reviews.

Newsfile Xtra

ONE DAY MY PRINTS WILL COME

This year there has been an interesting set of stories about so-called yeti encounters, the first of them which appeared in *The Daily Mail*. The most satisfying thing from my point of view is that, possibly for the first time in my life, neither *The Daily Mail* nor *The Sun* misquoted me.

I still stand basically by what I said and do not believe that these foot prints are anything to do with the Yeti.

I wrote to them:

Some of the pictures appear to show a degree of bifurcation, which might suggest that they are from an ungulate. What we know about the yeti would suggest that its foot prints would be in line with those other bipeds. The fact that the tracks appear to be in a straight line would suggest an animal that has evolved to ease its way along narrow mountain ledges. I would suggest a takin, which is after all the national animal of Bhutan.

And after viewing the video from which the stills were taken:

I think that part of what the cameraman said proves my point. He describes the part of the mountain where the tracks were found as being pretty well vertical. In my opinion, the only animal which could have negotiated such a terrain would be a mountain goat, or a goat antelope like the takin. I think that the chances of this set of prints from being anything more interesting is negligible, and it is certainly not a bipedal higher primate. The centre of gravity for such an animal would mean that it's just wouldn't be able to venture up a mountain like that.

And I am quoted as saying:

"Jon Downes, director of the Centre for Fortean Zoology, said the slope is so steep that only an animal like a mountain goat would have been able to negotiate it. He said: 'I think that the chances of these prints being from anything more interesting are negligible and it is certainly not a bipedal higher primate. The centre of gravity for such an animal would mean it just wouldn't be able to venture up a mountain like that.'"

Which isn't too bad.

The story has been picked up by all sorts of people http://tinyurl.com/zh3t3pm and so far none of the papers have written something portraying me as a complete weirdo. I am a complete weirdo, of course, but I don't need Her Majesty's Press to point out the fact!

The takin is a goat-antelope found in the eastern Himalayas, and is most closely related to sheep, although it has a peculiar to muskox. There are four sub species and one of them is a national animal of Bhutan which is – peculiarly, but not partially surprisingly - where the prints were found. One does wonder where the native guides were not merely telling the photographer what he wanted to hear.

The second yeti story, also from *The Daily Mail*, is far less substantial. On the 6th February they ran a story suggesting a yeti had been photographed in the Spanish ski resort at Formigal. There are a number of reports of man beasts from the Pyrenees, although I have my doubts whether these things could actually be flesh and blood, animals. I am even less convinced by the photograph which appears to show a bloke in a monkey suit. I won't even open the can of worms which is opened every time somebody tries to convince Richard Freeman that there are white yetis. Life is just too short for the out of control rant which would doubtlessly transpire!

A few days after I wrote the above, a website called The Lad's Bible printed a story confirming that the yeti sighting was nothing more than a marketing ploy by the company that owns and runs the ski resort. Now, why are we not surprised?

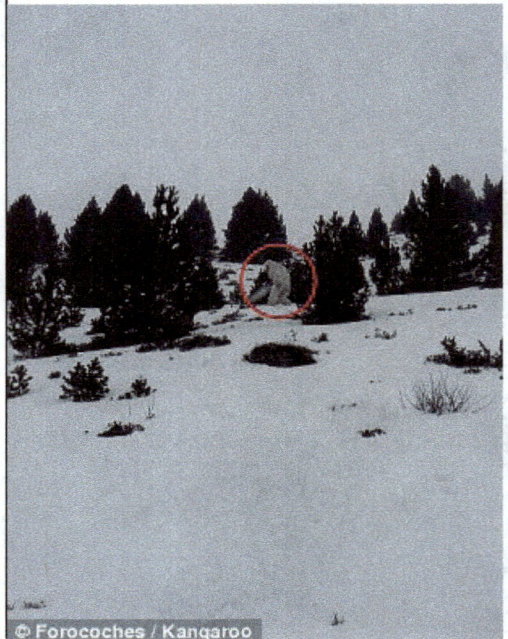

© Forocoches / Kangaroo

Newsfile Xtra

THE SUSSEX BIGFOOT

As a naturalist and self professed zoologist (I don't have a degree but as the word zoology comes from the Greek meaning the study of animals, I see no problem with proclaiming myself as such) I am quite prepared to accept the very strong possibility that there are unknown species of higher primate in the remoter parts of Asia. I am also prepared to accept that one or more of these unknown species of higher primate maybe bipedal, which would make them a fairly good match for the traditional accountants of yeti and almasty. I have always been less convinced by the North American man beast reports but during my various trips to Texas, meetings with Chester Moore Jr. and Ken Gerhard have persuaded me that there is substance behind some, at least, of the stories of Bigfoot and skunk apes.

Other reports from around the world, however are – to my mind at least – far more problematical. I have always been interested in the reports of Bigfoot –like creatures from the UK, even before I saw one myself.

My encounter, by the way, in case you didn't know, took place in January 2003 at Bolam Lake in Northumberland and is covered in some detail in my 2004 book *Monster Hunter*. CFZ Press published two books by Nick Redfern dealing with British 'Bigfoot' sightings, which are few and far between, despite what the internet would have us believe.

As the nation was preparing for the 97th anniversary of Armistice Day, and the newspapers prepared themselves to vilify poor Jeremy Corbin over the events in any way that they could, an interesting news item

caught our eye.

Huffington Post reported:

"A British dog walker photographed a Yeti over the weekend… in Sussex. The creature, dubbed the 'British Bigfoot,' was first spotted by Caroline Toms' border collie, Ash, who was startled by the ape-like creature while walking in woods. "It all happened so fast," she said. "Ash started acting a little bit bemused and barking. Then she quick as a flash shot off into the undergrowth - then I saw this big black thing flash out in front of me. Toms continued: "I only had my camera out because I was taking pictures of Ash playing. She came running back quick-as-a-flash though. I don't know what it was, but when I had a closer look at the pictures, it certainly does look like Bigfoot to me."

I have to say that the photograph was less then convincing. Corinna pointed out that it was mildly reminiscent of a photograph that we picked up on one of our trail cameras in a local wood, which appeared to show a female hiker with her trousers round her ankles presumably having answered a call of nature.

We are not racist, sexist, or anything elseist here at the CFZ and the fact that she was quite a cute blonde of a certain age does not prejudice our view of Ms Tom's testimony, but the fact that the report continued:

"I cannot be absolutely certain," she said. "It was just luck I had the camera out. It was so big and massive, I don't know if it could be anything else."

… Did not really give one any great hopes that an event of scientific or Fortean significance had taken place.

On Tuesday morning, knowing nothing of these events I was woken at about half past 7 by a flamboyant young man from the BBC wishing me to comment on some breakfast show or other. I decided to do the interview and stayed awake for half an hour or so I wish that I hadn't, because after the spooky music and the chatty nonsense of the presenter I was only on air for about a minute before I was cut off mid-sentence so they could go to another story. I then went back to sleep and overslept.

Last week I was reading about the furore currently surrounding events witnessed by our friend Adam Davis and others in – I believe – Oregon when I came across the following passage:

"There was a time when the majority of Bigfoot Researchers thought that the Sasquatch/Forest People were descendants of Gigantopithecus (i.e., Latin for Giant Mountain Ape). They thought that they were dumb giant mountain apes that could be chased through the woods and could be captured and/or killed. Wait!!! The majority of Bigfoot Researchers still think that way in spite of the fact that way too much data says otherwise. LOL!!! Yeah, right!!! Bigfoot is a dumb giant mountain ape. Then some silly Habituators came along and hypothesized that the Bigfoot species appear to be Forest People with special abilities, just like the North American Indians have been saying for thousands of years (i.e., The Bigfoot/ Forest People have special abilities such a Mind Speak, Zapping, Cloaking, Affiliation with Orb activity, Use of portals, etc.). It turns out that the North American Indians have been right all along"

I would agree that there are BHN reports from around the world which cannot be explained using a purely zoological model. Certainly my experience at Bolam comes into that category but was what Ms Tom's encountered in Sussex a zooform phenomenon of this type, or was it some bloke caught short behind a bush?

The day after my rude awakening from the BBC, the *Daily Mail:*

"the sighting was supposedly Kenny Matthews, 25 a rugby fan who was playing hide and seek with his two children, his girlfriend Lucy Buzzard – who works as a bar maid at the Fontwell Park race course in West Sussex said her boyfriend of 7 years, Kenny Matthews, was 'Bigfoot'.

She said: "the whole thing is hilarious, its my partner playing hide and seek with our two kids. The image looks like a man crouching down and hiding[…] Kenny had his hood up at the time and was wearing boots and his work clothes, dark trousers with a reflective patch, which seems to be in the photo."

But once again, for the second time this issue, it turned out to be a hoax perpetrated by an advertising agency. **http://tinyurl.com/h3gfb96**

I am afraid, that this once again is an example the power of the media. A succession of credulous reality TV shows have made Bigfoot into a media celebrity, and – unfortunately – the human race has always had a tendency, like Captain Ahab – to "smell land where there is no land" or in this case to see Bigfoot when there is

Carl Marshall's Column

Of all the alleged accounts the world over of anomalous, bipedal ape-like creatures; from the massive types such as the Himalayan yeti or bigfoot from the Pacific northwest, to the anthropomorphic almasty of Russia; none has quite the scientific credibility, and likely chance of discovery as that of the *orang pendek* or "short man" of Sumatra!

Reports vary slightly concerning this alleged creature but witnesses report some characteristics consistently. Typically, it is considered to be a small, 80cm to 150cm, bipedal primate, covered in greyish red to brown, or occasionally black fur. It is also said to have a mane of dark hair that runs from the head down the upper portion of its back. When walking, its gate is often described as making the animal look "low on its feet"! If it exists, it is likely to be predominantly herbivorous, feeding mainly on young shoots and fruit; a favorite is said to be the foul smelling durian fruit *Durio sp.*, which is also a favorite of both orang utan species *Pongo pygmaeus* and *P. abelii*! However, there are a few old reports of carnivorous activities such as that reported to Dr. Edward Jacobson in 1915 at *Sioelak Deras* where the indigenous population informed him that in the days when rhinoceroses were more common in *Kerinci*, *orang pendeks* used to be seen more frequently!

They claimed that they once caught the rhinos in pits which occasionally trapped them with their feet upturned, and one sometimes saw an *orang pendek* sat upon the rhinoceros's stomach, consuming the flesh! They are also sometimes said to eat small mammals and reptiles. *Orang pendeks* are likely also to be insectivorous, feeding on larvae in rotten tree stumps as sightings of such activities have been consistently reported. Dr. Jacobson, mentioned earlier, also claimed that in 1916, while camped near the base of *Boekit Kaba Mountain*, some scouts informed him they had recently observed an *orang pendek* breaking open a rotten log looking for insect larvae.

Therefore if all these activities are attributed a single species, *orang pendek* must be considered an opportunist

The Orang Pendek - Myth, Legend, or Reality? Part one

omnivore! They generally seem to be solitary animals, however there are a few reports of these alleged creatures being observed in small groups.

Word of this strange animal may have first left Indonesia as early as 1292 when Marco Polo, then the first westerner to visit the island, wrote of the province of Lambri (possibly Jambi), and stated:

"Now you must know that in this kingdom of Lambri there are men with tails; these tails are of a palm in length, and have no hair on them. These people live in the mountains and are a kind of wild men. Their tails are about the thickness of a dog. There are also plenty of unicorns [Rhinoceros] *in that country, and abundance of game in birds and beasts."*

This report may have little in connection with the *orang pendek* but is curious nonetheless! Marco Polo had not personally observed these so called "tails" and was certainly relying on already existing native myths and legends in this area for information. The notion that neighbouring tribes, are more primitive than ones own people, and thus being more animal-like, exists the world over; and the "tail" reported to Marco Polo was probably simply a representation of savagery! Marco Polo also described a strange ape-man dance performed at certain ceremonies.

Nothing more of Sumatra's "small man" reached the west until 1784, when English orientalist William Marsden

DCL FRS, who also edited an English edition of Marco Polo in 1818, *Travels of Marco Polo*, mentions an account in his own publication, *The History of Sumatra: containing an account of the government, laws, customs and manners of the native inhabitants, with a description of the natural productions, and a relation of the political state of that island.* In the book Marsden recalls, in as much detail as possible, reports of a species or race of small, yet robust hairy wildmen in southern Sumatra, that he compared to the orang utans of Borneo!

The first Dutch traders arrived in Indonesia during the late 16th century, along with the *Dutch East India Company*. By the 19th Century much of Indonesia had become a group of Dutch colonies and remained as such until after the second world war. Due to the long tradition of European settlers in Sumatra, there are many reports of *orang pendeks* by westerners during the early years of the 20th century.

As the names orang pendek and orang utan both start with the Malay word "orang" an incredulous reader might be tempted to suggest orang pendek is simply an orang utan, however this is not necessarily so! Firstly, there are no known populations of orang utans in this part of Sumatra and secondly, the name "orangutan" is derived from the Malay and Indonesian words orang meaning "person or "man" (as in human and not the male sex) or even "man-like creature" and "hutan" meaning "forest", thus "person of the forest". Originally, "orang

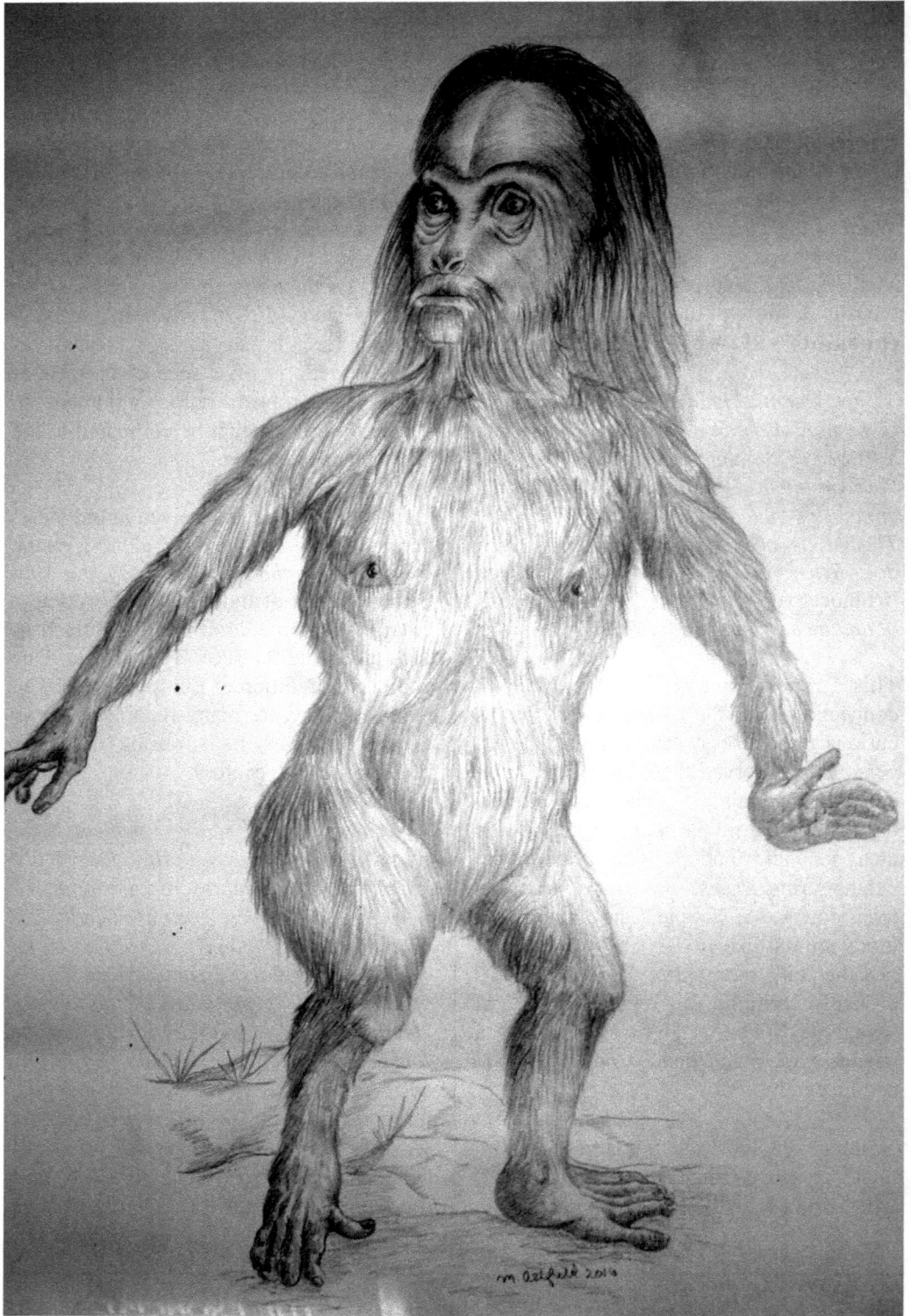

m aulfield 2010

*hutan" was not used to refer to apes, but to forest dwelling humans! The name is not even specific to anthropoids as in Borneo the proboscis monkey (*Nasalis larvatus*) the male of which has a proboscis 10.2 cm long, is actually known as orang belanda, meaning "Dutchman", as to the Malays the Dutch settlers, or for that matter any European, have more pronounced noses than the indigenous populations who have rather short noses. Incidentally, orang utang (a frequent misspelling) actually means "man in debt"!*

Some selected 20th century reports

In 1918, the Governor of Sumatra, L. C. Westenek, wrote of an encounter that took place in 1910.

"A young boy from Padang employed as an overseer by Mr. van H-- had to stake the boundaries of a piece of land for which a long lease had been applied. One day he took several workers into the virgin forest on the Barissan Mountains near Loeboek Salasik. Suddenly he saw, some 15m away, a large creature, low on its feet, which ran like a man... it was very hairy and was not an orangutan; but its face was not like an ordinary man's. It silently and gravely gave the man a disagreeable stare and ran calmly away. The workers ran faster in the opposite direction. The overseer remained where he stood, quite dumbfounded, and when he returned to camp he set down in writing what he had seen. His little note is

in my possession."

L. C. Westenek recorded another encounter from 1917. While in the forests at the base of *Boekit Kaba*, Mr Oostingh; the owner of a coffee plantation at *Dataran*, claimed he saw an unusual and hairy human-like animal, sitting on the ground about 30ft away from him.

" His body was as large as a medium sized native's and he had thick square shoulders, not sloping at all. The colour was not brown, but looked like black earth, a sort of dusty black, more grey than black. He clearly noticed my presence. He did not so much as turn his head, but stood up on his feet: he seemed quite as tall as I (about 1.75m). Then I saw that it was not a man, and I started back, for I was not armed. The creature took several paces, without the least haste, and then, with his ludicrously long arm, grasped a sapling, which threatened to break under his weight, and quickly sprang into a tree, swinging in great leaps alternately from right to left... My chief impression was and still is 'what an enormously large beast!' It was not an orangutan; I had seen one of these large apes before at the Artis (the Amsterdam Zoo). It was more like a monstrously large siamang, but a siamang has long hair, and there was no doubt it had short hair. I did not see the face, for, indeed, it never once looked at me."

In May, 1927, a Dutch plantation employee, A.H.W. Cramer, who resides in *Kerinci*, reported observing an *orang pendek* from a distance of only about 10 metres. Cramer, who was with a native woodcutter at the time, reported that the creature had long hair and black skin, and ran away at high speed leaving small, human-like tracks.

Also in 1927, a sergeant major with the Topographical Service named Van Esch, while working in *Surulagun*, reported that he had witnessed a human-like figure approach him as he climbed down a steep cliff to collect water. Van Esch hid as the creature approached allowing him a clear view of the short yet robust animal which came to within 5 metres of his hiding place. The creature was about 1.3 metres tall and covered with brown fur, and had an upper torso approximately half a metre wide. It had a long head with a dark face and pronounced canine teeth. It became startled by a noise in the forest and ran away on two legs.

In October 1923, while quietly sat in hiding, hunting for wild swine on the island of *Pulau Rimau*, Dutch settler Mr van Herwaarden, saw a strange creature that he referred to as a *sedapa* (orang pendek is known as sedapa or sedapak in the south-eastern lowlands). He described the encounter thus…

"It must be a sedapa I thought. Hunters will understand the excitement that possessed me. At first, I merely watched and examined the beast which still clung motionless to the tree. While I kept my gun ready to fire, I tried to attract the sedapa's attention, by calling to it. It would not budge. What was I to do? I could not get

help to capture the beast. And time was running short, I was obliged to tackle it myself. I tried kicking the trunk of the tree, without the least result. I laid my gun on the ground and tried to get nearer the animal. I had hardly climbed 3 or 4 feet into the tree, when the body above me began to move. The creature lifted itself a little from the branch and leaned over the side so that I could then see its hair, its forehead, and a pair of eyes which stared at me. Its movements had at first been slow and cautious, but as soon as the sedapa saw me the whole situation changed. It became nervous and trembled all over its body. In order to see it better, I slid down onto the ground again.

The sedapa was also hairy on the front of its body; the colour there was a little lighter than on the back. The dark hair on its head fell to just below the shoulder blades, or even almost the waist. It was fairly thick and shaggy. The lower part of its face seemed to end in more of a point than a man's; this brown face was almost hairless, whilst its forehead seemed to be high rather than low. Its eyebrows were the same colour as its hair, and were very bushy. The eyes were frankly moving; they were of the darkest colour, very lively, and like human eyes. The nose was broad with fairly large nostrils, but in no way clumsy. It reminded me a little of black person's. Its lips were quite ordinary, but the width of its mouth was strikingly wide, when open. Its canines showed clearly from time to time as its mouth twitched nervously. They seemed fairly large to me, at all event they [the canines] were more developed than a

man's. The incisors were regular. The colour of the teeth were yellowish-white. Its chin was somewhat receding. For a moment, during a quick movement, I was able to see its right ear which was exactly like a little human ear. Its hands were slightly hairy on the back. Had it been standing, its arms would have reached to a little above its knees; they were therefore long, but its legs seemed to me rather short. I did not see its feet, but I did see some toes which were shaped in a very normal manner. This specimen was of the female sex and about 5 feet high.

There was nothing repulsive or ugly about its face, nor was it at all ape-like, although the quick, nervous movements of its eyes and mouth were very like those of a monkey in distress. I began to talk in a calm and friendly way to the sedapa, as if I were soothing a frightened dog, or horse; but it did not make much difference. When I raised my gun to the little female, I heard a plaintive "hu-hu" which was at once answered by similar echoes in the forest nearby.

I laid down my gun and climbed into the tree again. I had almost reached the foot of the bough, when the sedapa ran very fast out along the branch, which bent heavily, hung on to the end, and dropped a good 10 feet to the ground. I slid hastily back to the ground, but before I could reach my gun again, the beast was almost 30 yards away. It went on running and gave a sort of whistle. Many people may think me childish if I say that when I saw its flying hair in the sights, I did not pull the trigger.

I suddenly felt that I was going to commit murder! I lifted my gun to my shoulder again, but once more my courage failed me. As far as I could see its feet were broad and short; but the native legend that the sedapa runs with its heels foremost, is quite untrue."

Many researchers believe this description to be that of a separate Sumatran cryptid known locally as *orang kardil* or "tiny man" which, according to Richard Freeman's Sumatran guide, the late Sahar Dimus, is a totally distinct entity to the *orang pendek*! However, I personally find it unlikely that Sumatra conceals two extant species of large, unknown primates! It seems more plausible that the *orang kardil* exists today only as a genetic memory of a pygmy human that once existed in southeast Asia and now believed to have survived on the island of Flores up until as recently as 12,000 years ago!

The head of the *Indonesian Tiger Conservation Group*, Debbie Martyr MBE, began her research into the legend of *orang pendek* in the 1980s, and actually witnessed the creature herself whilst camped on the slopes of *Mount Karinci* at an altitude of about 11,150 feet. Her guide Jamruddin pointed out areas where the Sumatran rhinoceros *Dicerorhinus sumatrensis* and tiger *Panthera tigris sumatrae* could be found, and then, quite casually commented, that in the forested mountains east of *Gunung Tujuh* (mountain of seven lakes) *orang pendeks* were occasionally seen! When Debbie made a skeptical comment Jamruddin

informed her he had seen the creature himself twice! Debbie, together with professional wildlife photographer Jeremy Holden, actively engaged in a fifteen year search for *orang pendek* funded by *Fauna and Flora International*. Jeremy used motion sensitive camera traps set up in remote locations in the *Kerinci Seblat National Park* but failed to capture the creature on film, though he claims he did catch a glimpse of what must have been an *orang pendek* as he climbed over a steep ridge in the jungle!

Adam Davies, along with Andrew Sanderson and Keith Townley, travelled to Sumatra in 2001 and found and cast tracks believed to be those of an *orang pendek* as well as collecting hair samples in the *Kerinci district.*

Primate biologist, Dr. David Chivers of *Cambridge University*, compared the casts with those from known primates and other animals found in Sumatra and stated: *"The cast of the footprint taken was definitely an ape with a unique blend of features from gibbons, orangutans, chimpanzees and humans. From further examination the print did not match any known primate species and can conclude that this points towards there being a large unknown primate in the forests of Sumatra."*

Dr. Hans Brunner, a world renowned authority on mammalian hair, compared the samples brought back by Davis *et al* to those of other primates and local animals and concluded that they originated from a previously undocumented primate species!

The Batutut

Borneo's answer to *orang pendek* is known as the *batutut* which, according to Richard Freeman's informative book on the subject *Orang Pendek: Sumatra's Forgotten Ape*, is an onomatopoeic word imitating the creatures cry. The *batatut* is about 1.2 - 1.5 metres tall, nocturnal, and afraid of fire! Unlike the *orang pendek* the *batutut* is supposed to attack and kill people, and consume their livers. This is likely to be a folkloric exaggeration!

In 1970, while on an expedition in Sabah, Malaysian Borneo, British zoologist John MacKinnon, who would later become world renowned for his amazing discovery of new mammals in Vietnam during the 1990's, found many short and unusually broad human looking tracks. MacKinnon reported:

"I stopped dead. My skin crept and I felt a strong desire to head home... Farther ahead I saw tracks and went to examine them. I found two dozen footprints in all. I was uneasy when I found them, and didn't want to follow them and find out what was at the end of the trail. I knew that no animal we know about could make those tracks. Without deliberately avoiding the area I realised I never went back to that place in the following months of my studies."

The lack of investigation on this professional zoologists part is a shame, as it wouldn't be at all surprising if Borneo also conceals a species similar to the *orang pendek* of Sumatra, and given the rapid deforestation rate of Sumatra's primary rainforests, Borneo could well end up a final stronghold for this alleged creature!

I travelled to northern Borneo in 2013 with my friend and colleague Andrew Jackson from the *Stratford upon Avon Butterfly Farm*, (see *Animals & Men* 51) to study both the countries diverse local wildlife and to investigate any cryptozoological conundrums we could seek to out along the way. Richard Freeman, the Zoological Director of the *Centre for Fortean Zoology*, informed us before we embarked on our trip that the Bornean *batatut* had never been properly studied; so whenever possible we asked after this cryptid. Unfortunately, we found very little evidence of the *batatut* in the locations we were investigating. In fact, while questioning local guides on the *Kinabatangan River* in Sabah we had more luck when we referred to the creature as *orang pendek* as apposed to the onomatopoeic name *batatut*!

Borneo is the third largest island in the world and the largest island in Asia, and the chances of us finding evidence on this preliminary expedition were extremely unlikely, especially considering we had very few reliable locations to start our investigation, and were depending entirely on seeking out good quality and sensible reports along the way. All was not lost however, as we did hear of some other interesting Bornean cryptids such as an alleged 35ft saltwater crocodile from *Lok Batik* in Sabah reported to us by soldier, 10 metre+ pythons from *Ulu Kamanis*, giant melanistic orang utans from *Kilimantan* and hitherto undocumented luminous birds supposedly viewed several times in natural clearings in the deep forests of *Ulu Kamanis*!

Andrew and I intend to return to Borneo in the near future to continue our search for the batutut and also to investigate the claims of the giant reticulated pythons of *Ulu Kamanis*!

The Orang kardil, the Ebu Gugo, and the "Hobbits" of Flores

As mentioned earlier, the *orang pendek* is not the only cryptozoological primate said to exist in Sumatra. In fact, tales of small bipedal primates have actually been shared across much of southeast Asia for centuries! From the island of Flores, the Nage people tell stories of the *ebu gogo*; said to be a tiny race of three foot tall, primitive, and gluttonous human-like creatures, covered in hair; though slightly less hairy than the alleged *orang pendek*! And considered nuisances by the ancestors of the Nage, who eventually trapped most of them in a cave in central Flores named *Lia Ula* and lit fires at the entrance choking the creatures to death inside, with only a few individuals supposedly escaping the alleged massacre! Through folklore, this story has become widespread, with similar tales told of the *nittaewo* of Ceylon (now Sri Lanka). According to the legend the *nittaewo* once inhabited a region from *Panawa Pattu* in the east to the *Kattaragama Hills* in the west, and were a persistent nuisance to the original aboriginal people, the Veddha. In 1886, British explorer Hugh Nevill was told the story by a Sinhalese hunter of small ape-like creatures from one of the last Veddhas in *Leanama* in the southeast of the country. The hunter claimed the creatures slept in caves or in shelters they created in trees with leaves as roofs. They had no fire, and hunted small animals such as deer which they disembowelled and consumed raw, an activity which disgusted the Veddhas. They used crude stone tools and talked in a strange twittering language that only a few Veddhas could understand.

The *nattaewo* allegedly moved around the deep forests in troops of between ten and forty. Finally, sometime in the late 18th Century the Veddhas cornered their hairy adversaries in a cave and built a huge bonfire of brushwood at the entrance. Sounds familiar! The fire burned for three days supposedly smoking all the *nittaewo* to death. Unfortunately, the Veddhas of the area also became extinct and the exact location of the cave was lost forever! The story was independently corroborated by Fredrick Lewis, who, while exploring eastern *Uva* and *Panawa Pattu*, met a family at *Salavai* whose grandfather was a Veddha.

The old man, Dissan Hamy, told Lewis (who had not heard of Nevill's account) an old story about a race of hairy pygmies called the *nittaewo*. Hamy claimed they had been exterminated some five generations ago! He also claimed his grandfather had taken part in the burning out of the cave which suffocated the creatures cornered inside. These suggestive tales may have a kernel of truth about them and may simply be modern interpretations, passed on through word of mouth throughout southeast Asia; a sort of garbled description of an original event that could have actually taken place thousands of years before. It may not even involve an actual fire, just the knowledge of remains of small human-like creatures associated with mountains and cave systems!

The "Hobbit" of Flores, or *Homo floresiensis* (Flores Man), is a likely extinct species of **Hominan** widely excepted to be of the genus *Homo*, discovered in 2003 on the island of Flores in Indonesia in southeast Asia. Partial skeletons of nine individuals have been recovered, including one complete skull referred to as "LB1". These remains have been the subject of intense paleontological research to ascertain whether they truly represent a species distinct from *Homo sapiens*. *H. floresiensis* is remarkable for its small body and brain size and for its survival until relatively recent times; possibly as recently as 13,000 years ago!

Recovered alongside the skeletal remains were relatively advanced stone tools displaying archaeological sequences ranging from 94,000 to 13,000 years ago! An analytical link can easily be formed bridging the discovery of *H. floresiensis* and the *ebu gogo* legends told on Flores, and also to a certain extent the tales of the *nittaewo* once told in Sri Lanka! The discoverers, Mike Morwood and colleagues proposed that a variety of features, both primitive and derived, identify these individuals as a new distinct species, *H. floresiensis* within the taxonomic tribe **Hominini**; which includes all species that are more closely related to humans than to chimpanzees. The discoverers also proposed that *H. floresiensis* lived contemporaneously with modern humans on Flores! Doubts that the bones constitute a new species were voiced by Indonesian Anthropologist Teuku Jacob who suggested that the skull of LB1 was that of a microcephalic modern human.

However two studies carried out by paleoneurologist Dean Falk and her colleagues (2005, 2007) rejected this theory. Falk *et al.* (2005) has been rejected by Martin *et al.* (2006) and Jacob *et al.* (2006), but defended by Morwood (2005) and Argue, Donlon *et al.* (2006). Two orthopedic researches published in 2007 reported evidence to support species status for *H. floresiensis*. A study of three fragments of carpal bones (wrist bones) concluded there were similarities to the carpal bones of a chimpanzee or an early hominin such as *Australopithicus* and differs from the bones of modern humans.

Is The Orang Pendek Real?

This naturally poses the questions, could the *orang kardil* and *H. floresiensis* be one and the same species and could the *orang kardil* still survive somewhere isolated in Sumatra today?

Of course even today, this hypothesis is by no means impossible, however I think it is extremely unlikely! Taking modern reports into account, a new undiscovered species of great ape seems much more plausible than a surviving undiscovered **hominan**!

Personally, I believe a more likely explanation is the *orang kardil* and *ebu gogo* legends are both innate memories of *H. floresiensis*, past on generation after generation; from island to island, throughout southeast Asia and works on the same principal as Chinese whispers, which would account for some of the regional variations of reported sightings!

So how does all this fit in with the modern *orang pendek* sightings? it might at first seem that the *orang pendek* is simply an extension of the mythology of the *orang kardil*. However, due to contemporary sightings by indigenous peoples, professional naturalists and scientists alike, it seems more plausible that *orang pendek* is actually a real, flesh and blood primate, completely distinct from both the *orang kardil* and *ebu gogo* legends; and utterly independent from recent paleontological remains discovered on Flores!

It is completelya possible the *orang pendek* exists! And part two of this article will be focusing on more recent reports and theories, and will attempt to propose an identity for this mysterious animal believed to be a primate.

Orang Pendek Illustrations by Maureen Ashfield

Sources and Suggested Further Reading

- **Allen, Benedict**: *Hunting The Gugu: In Search Of The Lost Ape-Men Of Sumatra*, 1989, Macmillan
- **Eberhart, George, M**: *Mysterious Creatures, A Guide To Cryptozoology, vol 2*, 2015, CFZ Press
- **Forth, Gregory**: *Images Of Wildmen In Southeast Asia*, 2009, Routledge
- **Freeman, Richard**: *Orang Pendek, Sumatra's Forgotten Ape*, 2007, CFZ Press
- **Heuvelmans, Bernard**: *On The Track Of Unknown Animals*, 1958, Hart-Davis
- **MacKinnon, John, Ramsey**: *In Search Of The Red Ape*, 1974, Holt, Rinehart and Winston
- **Marsden, William**: *The History Of Sumatra Containing An Account Of The Government, Laws, Customs And Manners Of The Native Inhabitants*, 1811, PBA
- **Wikipedia**, *The Free Encyclopedia*

Carl Marshall works at Stratford Butterfly Farm and is a fine field naturalist. Over the past couple of years he has become a very enthusiastic member of the CFZ, and his quasi-Fortean view of British natural history fits in perfectly with my own. He was, therefore, the perfect choice as a columnist for the brave new *Animals & Men*, and we are proud to have him aboard.

watcher of the skies

CORINNA DOWNES

Pakistan lifts ban on rare bird hunts

On the 22nd January this year, Pakistan's Supreme Court lifted a ban on hunting the houbara bustard, a rare bird, because its meat is prized among Arab sheikhs as an aphrodisiac.

People from the Gulf travel to Balochistan province every winter to kill the houbara bustard using hunting falcons. The issue of hunting came into limelight after a report in 2014 showed Saudi prince Fahd bin Sultan bin Abdul Aziz Al Saud killed over 2,100 houbara bustard in a cruel 21-day campaign in clear violation of his permit to hunt only 100 birds. Houbara bustard is listed in the Convention on Migratory Species of Wild Animals, also known as the Bonn Convention, and is declared as an endangered species by the International Union for Conservation of Nature.

SOURCE: http://tinyurl.com/gmeuz45

What makes this even more confusing is that the species is now completely different: MacQueen's bustard (*Chlamydotis macqueenii*) is a large bird in the bustard family. It was earlier included as a subspecies of the houbara bustard (*Chlamydotis undulata*) and sometimes known as the Asian houbara. The subspecies are geographically separated from the houbara found west of the Sinai Peninsula in North Africa with a population in the Canary Islands. MacQueen's bustard is found in the desert and steppe regions of Asia, east from the Sinai Peninsula extending across Kazhakstan east to Mongolia. These two species are the only members of the genus Chlamydotis. MacQueen's is a partial latitudinal migrant while the houbara bustard is more sedentary. In the 19th century, vagrants were found as far west of their range as Great Britain. Populations have decreased by 20 to 50% from 1984 to 2004 due mainly to hunting and land-use changes.

Europe's rarest seabird 'faces extinction'

The BBC reported on the 12th March that, according to new analysis, Europe's rarest seabird, the Balearic shearwater (*Puffinus mauretanicus*), will be extinct within 60 years. Scientists say that urgent action is required to stop the incidents of the bird being drowned in fishing lines and nets, according to findings published in the *Journal of Applied Ecology*.

The bird breeds in the Balearic Islands, sometimes stopping off in British waters as it migrates north. There are about 3,000 breeding pairs left, and research shows the global population is not sustainable in the long term. This seabird is classified as critically endangered on the IUCN Red List of species, and breeds on cliffs and small islets and lays only one egg.

Prof Tim Guilford of the Department of Zoology at the University of Oxford is co-researcher on the study, and he told BBC News: "The survival of adults from one year to the next and especially of young adults is much lower than we thought. The species is unsustainable - it is on the road to extinction."

Estimates suggest about half of deaths in adult birds are due to accidental capture in fishing lines and nets. Changes such as setting fishing gear at night when the bird does not dive "could make a massive difference", said Prof

Guilford. He added, "The science shows just how serious the problem is, but also that there is a technically simple solution - the setting of demersal long-lines at night."

The Balearic shearwater is one of the rarest seabirds in the world.

SOURCE: **http://tinyurl.com/jsvq26t**

Game & Wildlife
CONSERVATION TRUST

Results revealed for 2016 Big Farmland Bird Count

It was reported on 26th March that threatened species are among the most commonly-seen birds in East Anglia, according to the results of the 2016 Big Farmland Bird Count (BFBC).

The Game and Wildlife Conservation Trust (GWCT) launched the annual count in 2014 to highlight the positive work done by farmers and gamekeepers to help reverse the decline in farmland bird numbers, and this year, nearly 1,000 farmers across the country spotted 130 different species, which is the highest total so far. They included 25 species from the Red List for Birds of Conservation Concern, with six appearing among the top 25 most commonly-seen species: Fieldfares, house sparrows, starlings, yellowhammers, song thrushes and skylarks.

More than 100 different species of birds were spotted in Norfolk, across a huge area of nearly 63,000ha of land. Lapwings, starlings and linnets were in the top 10 most abundant birds seen in the county, spotted along with 17 other Red List species.

In Suffolk, fieldfares, yellowhammers and starlings were among the top 10 most abundant species seen on more than 22,000ha of land.

The GWCT's head of development and training, Jim Egan, said: "Despite the horrible weather at the start of the count week, we've nearly doubled the total number of participants since the first year. It really does show that farmers have a long-term commitment to conservation management." Nationally, the most commonly seen species were blackbirds and woodpigeons, seen by more than 80pc of participants. The five most abundant birds seen were woodpigeons, starlings, rooks, fieldfares and lapwings – which made up 40pc of the total number of birds recorded. Farms taking part were of varying sizes and type, with more than 60pc in some form of agri-environment scheme, and more than half providing some form of extra seed feed for birds, either through growing wild bird seed mixes, or by providing additional grain through scatter feeding or via hoppers.

SOURCE: **http://tinyurl.com/hbhsgmu**

Rare and/or out of place bird sightings from around the world

Egyptian vulture

It was reported on March 6th, that a very rare species of vulture, the Egyptian vulture (*Neophron percnopterus*) which is on the verge of extinction, was sighted in Ganjam District of Odisha, an eastern Indian state on the Bay of Bengal.

Dr Jyoti Prasad Pattnaik, founder of Berhampur Birds, a self-trained bird enthusiast sighted this rare bird in the area of Bhetanai, while he was recording the movements of the endangered blackbucks in the region. The number of vultures are going down day by day due to the use of pain killer drug Diclofenac in sick animals and these birds which survive on the carcases of the dead sick animals succumb to the ill effects of the drug.

SOURCE: http://tinyurl.com/zsuxlhr

4,000 miles from home, rare giant pelican appears on Sanibel Island

A great white pelican has appeared at the J.N. "Ding" Darling National Wildlife Refuge on Sanibel Island, which is around 4,000 miles from its home range. It is also the first time the Old World species has been recorded in North America.

Common brown pelicans that are seen in southwest Florida can be around 11 pounds with a 6½ foot wingspan, and the American white pelican is even larger, but the great white pelican can be around 33 pounds with nearly a 12ft wingspan. Because the bird's facial skin is yellowish, it's likely a female. Males have pinker skin. It appeared healthy and getting along well with the other pelicans nearby, said Fort Myers paediatrician and passionate birder Jose Padilla-Lopez, who quickly hustled out to see and photograph it.

However, there was debate on social media about whether this pelican is, in fact, countable as a rare visitor. It could be an escapee from a zoo or private owner, though this bird has no identifying tags or bands. Padilla-Lopez said, "Some are dismissing it off the bat as an escapee, others — rightly so in my view — want to give it time and consider that it may be a vagrant (and countable) bird."

A spokesperson from J.N. "Ding" Darling National Wildlife Refuge said, "We know national wildlife refuges are critical places for wildlife and this justifies why we need them – safe places for all species to rest and feed. Even if this is just a stopover for a visiting rare bird like the great white pelican, refuges are critical."

SOURCE: http://tinyurl.com/zhbuscp

Himalayan griffon spotted in Goa

On 12th February it was reported that Mandar Bhagat and Omkar Dharwadkar of the Goa Bird Conservation Network (GBCN) said they spotted a rare Himalayan griffon, also knows as the Himalayan vulture (*Gyps himalayensis*), in Cacora village.

"Notes taken from the field and photographs of the bird taken were sent to several expert ornithologists across the country to confirm the species and our suspicions were correct. It is indeed the Himalayan griffon," said Mr. Dharwadkar, who was the first to spot the avian.

According to the GBCN, the Himalayan griffon was previously believed to belong to the upper Himalayas and was presumed to stray till the Gangetic plains at the most. In 2013, however, "an exhausted juvenile" was rescued in Thrissur district of Kerala. In the same year, multiple sightings of the species were also reported from Bangalore in Karnataka and Kakinada in Andhra Pradesh. Earlier this year, the same species was reportedly spotted in Kaiga in Karnataka, the network of avid birdwatchers said.

"Himalayan griffons do not breed in the first three years, and hence juvenile birds of the species do not remain in breeding grounds to avoid competition. Such long-distance straying from home territory also points towards a lack of navigational experience in immature birds. All individuals of the species previously reported as sighted from south India, including the one spotted in Goa, are the immature ones. With this, the list of birds of Goa officially stands at 460 species, of which 14 additions were made in the last three years alone," said Pronoy Baidya, a reviewer for *eBird*, an online programme that crowdsources information from birdwatchers.

Ornithologists said there was a time when vultures could be seen in large numbers in Goa, especially the white-rumped vulture (*Gyps bengalensis*) and the Indian vulture (*Gyps indicus*). However, exposure to large doses of diclofenac has spelt doom for these birds as they feed on carcasses of cattle given this painkiller. Today, vultures can be said to be locally extinct in Goa, they added.

SOURCE: http://tinyurl.com/jts9d47

Ring-billed gull in Strathclyde Park

On 17th February, it was reported that a rare visitor, a ring-billed gull, was seen at Strathclyde Park. This bird is the most common member of its species in the United States, and is now starting to travel across the Atlantic. This particular bird is one of only two who are known to have set up home in the Scotland, with the another found in Dingwall.

Bird watcher, John Nadin, travelled from Dunfermline to photograph it for himself, and said, "While I was at Strathclyde Park last week I met at least five other bird watchers who were there to see and photograph the returning North American gull, so it is becoming a popular bird with local and visiting bird watchers alike. It is still a rare bird in the British Isles, currently there are two ring-billed gulls in Scotland, the other is in Dingwall and it has returned to the same site for several years now. Another notable ring-billed gull returned to Stromness in Orkney for over 20 years.

"The first ring-billed gull ever to be seen/ identified in the UK was at Blackpill Beach in Glamorgan, Wales, in 1973, with the first recorded in Scotland on the Ythan Estuary three years later.

"It has increased a great deal in the UK since these early records mainly due to increases in the North American breeding population, also there are more birdwatchers, with better optics and better field guides, as well as local, national and Internet-based information services that report unusual birds."

SOURCE: http://tinyurl.com/glt5hto

Migratory duck makes rare visit to Singapore

A rare visitor, a male northern pintail, flew into Singapore's Sungei Buloh Wetland Reserve. A breed of duck, it was seen feeding among a group of egrets in the reserve. The last time this species of bird was sighted here was in 1992 in Senoko South. This is the first time that the species has been seen in the wetland reserve, said the National Parks Board (NParks).

Mr Wong Tuan Wah, director of conservation at NParks, said the migratory

bird is a "very rare visitor" to Singapore. He added: "We are excited that the northern pintail has decided to stop over at Sungei Buloh Wetland Reserve - particularly in the week of World Wetlands Day. This shows that the wetland reserve is a significant stopover point for migratory birds."

According to conservation group BirdLife International, the bird is native to countries such as Britain and Switzerland. The omnivorous creature feeds on algae and amphibians, among other things. Its sighting comes as the avian migratory season is about to end.

Mr Alan OwYong, from the Nature Society's Bird Group, said, "The northern pintail migrates from North Asia to South Asia, stopping over in Indochina... I personally think that the cold snap south of China may have forced the birds to fly further south."

SOURCE: http://tinyurl.com/hz7bl5u

Rare bird reappears in Beijing after 70 years

Beijing News reported on the 4th February that a small flock of Jankowski's buntings (*Emberiza Jankowskii*) were spotted at Miyun Reservoir in Beijing, about 70 years after the last recorded sighting. Xing Chao, a student at Peking University and also a member of a Beijing birdwatching group, is reported to have found the endangered bird at the reservoir. It is the first time evidence of the endangered species has been found in Beijing since 1941, when traces of the *Emberiza Jankowskii* were spotted at the Summer Palace. This bird is a species in the Emberizidae family, and there are less than 1,000 left in the wild, and this

discovery has provided a unique opportunity to study the birds, of which little is known, the report said.

The bird normally lives in China, the Democratic People's Republic of Korea and Russia, where its natural habitats are temperate shrublands and grasslands. However, in the last four decades the species has rapidly declined in number, primarily because its habitat is increasingly being converted for use in agriculture, farming and forestry.

SOURCE: http://tinyurl.com/zkhnfks

Birders flock to Surrey farmland for rare Asian bird

According to the *Metro Vancouver*, in January birders from around the continent are taking flight to a patch of south Surrey farmland to see a striking little bird that belongs in Asia — the Siberian accentor.

"This particular bird is in an easily accessible area," George Clulow, president of B.C. Field Ornithologists, said in an interview Friday. "People can easily fly in, which is what they've been doing from all over North America. I was standing next to people from Massachusetts, California, Florida, Minnesota."

This bird is normally found in Southeast Asia and breeds across Siberia.

SOURCE: **http://tinyurl.com/zqe7rqn**

Rare bird captured on camera in Denbighshire

A snow bunting, a rare bird to our shores, was spotted by a St Asaph photographer around the 21st January. He was visiting one of Denbighshire's pebbled beaches when he spotted the small brown bird.

According to the RSPB the snow bunting breeds globally in places such as Scandinavia and Alaska and migrates further south in winter. A spokesman from the charity said: "They are a scarce breeding species in the UK, in Scotland, making them an Amber list species."

The RSPB splits birds into three categories of red, amber and green – amber is the second most critical category for birds found in the UK.

SOURCE: **http://tinyurl.com/j3phkcd**

Rare bird found in Höfn, East Iceland

"I had no idea what kind of bird we were looking at, it was so weird," says ornithologist Brynjúlfur Brynjúlfsson at the South East Iceland Bird Watching Centre. He is the first person to have spotted a dark-Sided flycatcher (*Muscicapa sibirica*) in Western Europe.

The species, according to Wikipedia, breeds in South-East Siberia west to beyond Lake Baikai as well as in Mongolia, China, North Korea and Japan.

Their wintering range includes India, Bangladesh, southern China, Taiwan, Sumatra, Java, Borneo and the Philippines. Vagrant birds have been previously recorded as far as Alaska and Bermuda.

SOURCE: http://tinyurl.com/gwqudf9

Rare sighting of goosanders in Ipswich

A pair of male and female Gooseanders were spotted in Christchurch Park in Ipswich on 17th January.

They are a member of the sawbill family, and Chris Courtney, leader of the RSPB Ipswich group, said: "You find them in upland rivers in northern Britain in the summer, that's where they breed. In the winter they disperse and go further afield looking for food. They are winter visitors to Suffolk."

He added that it was rare for the birds to visit Christchurch Park, with him only having seen one female two years ago, and two in the park before Christmas.

In recent years they have also been sighted in Lackford Lakes, in Bury St Edmunds, and Alton Water's conservation area.

SOURCE: http://tinyurl.com/z5zxf2j

Rare bird spotted in Gloucester brook

A penduline tit has only been recorded once in the Gloucestershire area since records began, but in January one was spotted on Horsbere Reserve, which was developed by the Environment Agency as part of the flood management scheme. The reserve is now owned and managed by Gloucester City Council, and is made up of a large wetland that diverts flood water away from properties and into the storage area at times of a flood, as well as being an important wildlife habitat.

Cllr Jim Porter, cabinet member for environment at Gloucester City Council said, "It is very exciting to have such a rare bird visit the reserve. Horsbere Reserve was only built a few years ago and already we have had visits from various wetland birds such as herons, kingfishers and egrets. It is good to see that an area designed for flood protection is also benefitting nature conservation." Mike Smart Honorary Chairman of the Gloucestershire Naturalists' Society, Gloucestershire Wildlife Trust trustee and local resident added: "This is an exceedingly rare bird only once ever before being recorded in Gloucestershire. The species has spectacular plumage and originates in central Europe; it gets its name from its habit of hanging upside down on bulrushes on which it feeds."

SOURCE: **http://tinyurl.com/jyntg2j**

Winter storms brings rare birds to British shores

Winter storms in January created an amazing "wreck" of one of the world's smallest seabirds, as high winds and rough waters brought more than 100 little auks ashore. They were nursed back to health before they could be released. The auks, about the same size as a starling, were appearing in gardens and vegetable patches when they should have been feeding on plankton well out in the North Sea. There were also reports of thousands of little auks passing by coastal headlands and piers as they headed towards shore looking for shelter from the gales. For the Scottish Society for the Prevention of Cruelty of Animals, the shock arrival of so many little auks meant a busy New Year for staff at its National Wildlife Rescue Centre in Fishcross. Centre manager Colin Seddon said: "We have just over 100 little auks in our care at the moment which have been caught out by the recent storms. Little auks breed in High Arctic areas such as Greenland and Iceland, so it is unusual to see them up close. It is not uncommon for little auks to be found in the North Sea over winter but they have been blown off course during the recent storms and are landing in areas up and down the county, predominately along the East Coast.

He added, "The little auks we have rescued were found weak and thin and would have had great difficulty taking off once grounded."

SOURCE: **http://tinyurl.com/j7ex4o9**

Bird species makes rare Scottish appearance

Also in January, a glossy ibis, a bird more commonly seen in southern Europe and Africa,

made a rare appearance in Scotland. It was spotted at the ruined Ormiclate Castle on South Uist in the Outer Hebrides, and its sighting came a few days after a swift was spotted in Scotland for only the fourth recorded time.

The bird was discovered at Thortonloch, near Dunbar, in East Lothian, on Hogmanay. The bird has the body of a curlew and the legs of a flamingo and gets its name from the iridescent sheen on its wings. It has become a regular winter visitor to the south of England, with young birds appearing in flocks from Spain.

SOURCE: **http://tinyurl.com/h9aua3u**

Crazy winter weather brings rare red-rumped swallow to British shores

Red-rumped swallows were spotted in Britain during late December last year. They are from the Mediterranean, but ended up spending Christmas on the North Norfolk coast at Wells-next-the-Sea. It should really have been south of the Sahara or even in India rather than Norfolk in the UK.

SOURCE: **http://tinyurl.com/z6sx83e**

Rare 'butcher bird' spotted at Surrey Wildlife Trust reserve in Camberley

An officer of Surrey Wildlife Trust (SWT) spotted a great grey shrike (*Lanius excubitor*) at Surrey Wildlife Trust's Poors Allotment nature reserve near Camberley in March. The bird of prey is a rare and elusive bird, which winters in the UK in small numbers.

"It's an incredibly rare bird and it's very

difficult to get a glimpse, so I was really lucky to witness it," said Mr Herd, the officer.

This shrike is not much bigger than a blackbird, and hunts small mammals, lizards and beetles and will even kill other birds as big as greenfinches. It then stores its catch in a bush or tree, to devour later.

"It's known as the butcher bird, because it has this unusual behaviour of keeping its prey in a makeshift larder," said Mr Herd "Sometimes, it even impales mammals or birds on a thorn for safekeeping."

Only 200 Great Grey Shrikes visit the UK every year between October and May, travelling from Europe, Asia and north Africa. Mr. Herd explained that Poors Allotment offers the perfect habitat for the species, a relatively quiet heathland with a good food source and plenty of high perches so the bird can look out for prey.

"SWT works hard to preserve this type of heathland habitat, which is vital for these birds," he added. "If we lost these habitats, the shrike would have nowhere to live in winter."

The few Great Grey Shrikes wintering in the UK will soon migrate back to their breeding grounds in Scandinavia. SWT has appealed for anyone who spots or photographs this rare bird to record the sighting at www.surreywildlifetrust.org/SBIC

SOURCE: http://tinyurl.com/jsxh8bq

New and rediscovered bird species

NEW SPECIES OF BIRD DISCOVERED IN INDIA AND CHINA BY INTERNATIONAL TEAM OF SCIENTISTS

On 20th January it was reported that a new species of bird had been discovered in northeastern India and adjacent parts of China by a team of scientists from Sweden, China, the U.S., India and Russia.

As described in the journal *Avian Research*, the bird has been named Himalayan forest thrush (*Zoothera salimalii*). The scientific name honours the great Indian ornithologist Sálim Ali, in recognition of his contributions to the development of Indian ornithology and nature conservation.

Pamela Rasmussen, of Michigan State University's Department of Integrative Biology and the MSU Musem, said that the discovery process for the Himalayan forest thrush began in 2009 when it was realized that what was considered a single species, the plain-backed thrush (*Zoothera mollissima*), was in fact two different species in northeastern India.

Rasmussen was part of the team, which was led by Per Alström of Uppsala University (Sweden). The scientists' attention was first caught by the fact that the plain-backed thrush in the coniferous and mixed forest had a rather musical song, whereas individuals found in the same area – on bare rocky ground above the treeline – had a much harsher, scratchier, unmusical song. "It was an exciting moment when the penny dropped, and we realized that the two different song types from plain-backed thrushes that we first heard in northeast India in 2009, and which were associated with different habitats at different elevations, were given by two different species," Alström said.

Rasmussen added, "At first we had no idea how or whether they differed morphologically. We were stunned to find that specimens in museums for over 150 years from the same parts of the Himalayas could readily be divided into two groups based on measurements and plumage."

"To an ornithologist, the Himalayan forest thrush sounds like Adele, while the alpine thrush sounds more like Rod Stewart," said study co-author and Wildlife Conservations Society associate Shashank Dalvi. DNA analyses suggested that these three species have been genetically separated for several million years.

enetic data also yielded an additional exciting find: Three museum specimens indicated the presence of yet another unnamed species in China, the Yunnan thrush. Future studies are required to confirm this.

New bird species are rarely discovered nowadays. In the last 15 years, on average approximately five new species have been discovered annually, mainly in South America. The Himalayan forest thrush is only the fourth new bird species described from India since 1949.

Rasmussen is tied for the third-highest number of birds discovered in the world since 1950 and is ranked first for birds discovered in Asia, and Alström is second for Asia in the same time period.

Additional scientists who contributed to the study include Chao Zhao (China), Jingzi Xu (Sweden), Shashank Dalvi (India), Tianlong Cai (China), Yuyan Guan (China), Ruiying Zhang (China), Mikhail Kalyakin (Russia), Fumin Lei (China) and Urban Olsson (Sweden).

SOURCE: http://tinyurl.com/jftocqg

Paragallinula: Ornithologists Describe New Bird Genus

An international team of ornithologists from Sweden and New Zealand has described a new genus for the lesser moorhen. A new genus Paragallinula is established for the lesser moorhen, which was previously assigned to the genus Gallinula. The team, headed by Dr George Sangster from Swedish Museum of Natural History, named the new genus Paragallinula.

"The generic name is derived from the Greek para (beside) and the Latin gallinula (a little hen or chicken)," Dr Sangster and co-authors wrote in a paper in the European *Journal of Taxonomy*.

"It denotes the resemblance of *P. angulata* to species of Gallinula but highlights that they are independent evolutionary lineages." According to the scientists, *Paragallinula angulata* is found in most of the African continent from Senegal and Gambia to Ethiopia, Namibia, Botswana and South Africa. George Sangster et al. 2015. A new genus for the Lesser Moorhen Gallinula angulata Sundevall, 1850 (Aves, Rallidae). European Journal of Taxonomy 153: 1 – 8; doi: 10.5852/ ejt.2015.153

SOURCE: http://tinyurl.com/hle48a3

Scientists Solve A Shag-adelic Bird Mystery

March 25, 2016: New Zealanders now get two endangered shags for the price of one.

Compared with kiwis and other island birds, New Zealand's dozen or so shag species have not been well studied, and for many years, scientists couldn't decide how many shag species lived in New Zealand. One species, the Stewart Island shag, was first described in 1845 by British zoologist George Robert Gray, and was listed as having two distinct populations, as well as bronze and pied morphs. While some ornithologists considered these one species, others dubbed them separate species or subspecies.

Recent research finally resolved the century-old debate. Scientists analyzed the birds' genes and concluded that the Stewart Island shag is actually two separate species: the Foveaux shag (*Leucocarbo stewarti*), which lives on both sides of the Foveaux Strait between Stewart Island and South Island, and the Otago shag (*Leucocarbo chalconotus*), which lives further to the east on South Island.

Nic Rawlence, lead author of a study published in the *Zoological Journal of the Linnean Society* in February, and a postdoctoral fellow at the University of Otago, said the team used ancient DNA from museum samples and sub-fossil remains to show that the two species have been separate for centuries. "The Otago species was spread around the entire eastern South Island," he says. "It never interacted with the Foveaux Shag." However, one of the species is under threat. Though they differ in appearance only slightly, the two species suffered disproportionately at the hands of humans.

Birds.B.M.Vol.XXVI.   *Pl.V.ᵃ*

Phalacrocorax stewarti

Rawlence estimates that as many as 99 percent of the Otago shag disappeared within 100 years of Polynesian settlement, and due to a genetic bottleneck the species now has almost no genetic variation. The Foveaux shag avoided this fate because it mostly lives on inaccessible rock islands just offshore where settlers could not disturb them.

The Stewart Island shag, which numbers around 5,000 birds, had already been considered vulnerable to extinction, but the split into two species creates new conservation priorities. Rawlence says initial estimates suggest roughly equal numbers of each, which

likely means both will be recategorized as endangered. The Otago shag's limited genetics, however, put it in worse shape. "The Foveaux Shag still has a lot of diversity, so as long as you stop predators from getting to the islands where they live, they should be okay," he says. "But the Otago shag needs tailored conservation." He suggested that establishing new colonies in parts of the species' former range may be an effective way to boost their population, although it would not solve the problem of their genetic diversity.

Meanwhile, both species benefit from efforts by the New Zealand Department of Conservation to preserve their coastal ecosystems. Because shags are commonly killed as fisheries bycatch, they benefit from efforts to mitigate that, too.

SOURCE: http://tinyurl.com/zwv6puv

For those of you not aware, as well as this column in *Animals & Men*, Corinna writes a daily Fortean bird blog which can be found as part of the CFZ Blog Network, but also as a stand alone site at:

http://cfzwatcheroftheskies.blogspot.com/

FURRY FAIRIES: THE STRANGE WORLD OF BOGEY-BEASTS

In an extract from his new book *The Hidden Folk: Are Poltergeists and Fairies Just the Same Thing?* (available now from CFZ Press), **SD Tucker** explores the strange world of bogey-beasts, and asks whether or not cryptozoology really ought to be reclaimed as a modern-day subgenre of fairy-lore.

In this year of 2016, the 400th anniversary of William Shakespeare's death, it seems

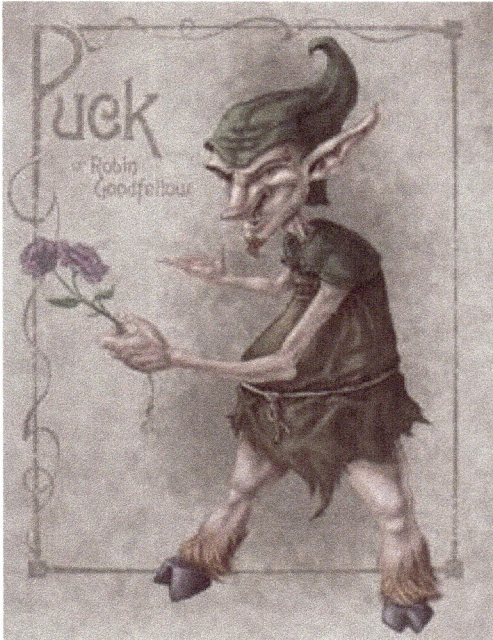

appropriate to open this article with a quote from perhaps his most well-loved play *A Midsummer Night's Dream* – namely, some lines uttered by the famous English fairy-Trickster Puck (AKA: Robin Goodfellow), in which he boasts about being able to shape-shift and appear in the form not just of a dwarfish humanoid, but also as various animals and glowing lights. Or, as Robin himself puts it:

Sometime a horse I'll be, sometime a hound,
A hog, a headless bear, sometime a fire,
And neigh, and bark, and grunt, and roar, and burn,
Like horse, hound, hog, bear, fire, at every-turn.

What was Shakespeare talking about? In spite of his prodigious powers of imagination, the Bard did not simply create these images from out of thin air, giving "to airy nothing a local habitation and a name", as he once put it. Instead, Shakespeare was picking up on genuine aspects of old English folklore of the time. It seems that strange animals – or animal-*like* beings, anyway – encountered in the lanes and by-ways of the countryside in centuries past were once thought of as being entities like Robin Goodfellow in disguise, as these lines clearly show.

These fairy-animals – whether conceived of literally as being fairies in shape-shifted form, or else as some separate and distinct beast-like class of elf – were known in Britain as the 'bogey-beasts' or 'barguests', this latter term probably meaning something like 'town-ghost'. They came in a variety of shapes – dogs, calves, donkeys, headless bears, even hairy ape-like beings called 'shug-monkeys' – and are thought of by most as having been purely legendary. This opinion may actually be wrong, however, for bogey-beasts are, apparently, still being seen, albeit under different names. Their shape-shifting nature might not be so commonly reported nowadays, but it seems obvious that certain legends surrounding well-known fortean figures like Black Dogs and uncatchable Alien Big Cats to some extent grew from these older stories about bogey-beasts.

A NEW FORM OF ZOOFORM?

Fairy-lore, in fact, is still very much alive in the modern world in disguised form, and I think it would be possible to identify three main different strands of it as being present in contemporary studies into matters fortean:

Firstly, there is the branch of modern fairy-lore which is most germane to my new book – namely the fact that, when what people would once have called fairy-phenomena occurs within a home or other building, we would now classify the events involved under the heading of 'poltergeistry'. Here, for example, is an old description of what the Italians used to term *foliots* – fairies – but what we ourselves would now surely call poltergeists. According to the sixteenth-century Italian writer Gerolamo Cardano, these *foliots* would "frequent forlorn houses" where they would then: " ... make strange noises in the night, howl sometimes pitifully, and then laugh again, cause great flame and sudden lights, fling stones, rattle chains, shave men [in their sleep], open doors and shut them, fling down platters, stools, chests, [and] sometimes appear in the likeness of hares, crows, black dogs, etc." Or, in other words, they would act like typical poltergeists are said to do today!

Secondly, as observed most famously by the French writer Jacques Vallee in his seminal book *Passport to Magonia*, there are numerous old legends concerning people supposedly being taken away to fairy-land or inside fairy-hills by the Good Folk. These tales, according to Vallee's famous theory, involving as they did such now-familiar motifs as 'missing-time', bright lights and interbreeding between human and non-human fairy-races, have simply been reconceptualised and reclothed by modern ufology in the whole dubious 'alien abduction' narrative.

Thirdly, though, there is the issue of fairy-animals, barguests, bogey-beasts and the like. When seen inside people's houses, strange animal-like apparitions have understandably been linked with poltergeists and there are even a few examples of such creatures allegedly being seen in conjunction with UFOs. However, when observed outside and 'in the wild', as it were, these particular classes of fairies seem nowadays to come under the general purview of cryptozoology, or 'the study of unknown animals'. Some cryptozoologists would no doubt object to this idea, pointing out that what they investigate are in fact real, flesh-and-blood animals, not fairies. Often, they are correct in this assertion. The coelacanth was no supernatural bogey-beast, and nor is the thylacine.

An alternative viewpoint, however, would have it that alongside such regular, zoological cryptids there exist in the world other strange, animal-like beings, apparently of a supernatural nature, which are not actually animals at all. As such, I think it would be eminently possible to argue that barguests, bogey-beasts and other such fairy-animals have been reconceptualised again in recent years in cryptozoology not only as

specific but limited types of animal-phantom like Black Dogs, but also under the more generic heading of 'zooform phenomena' – a term coined by the head of the CFZ, Jonathan Downes, in a 1993 essay. Here, Downes defined zooforms as being: "Apparitions which take the form of animals – usually living but which are not living things – at least in the way that we understand the term." This definition sounds not entirely unlike a good working description of some of the old bogey-beasts, of course ...

ANIMALS THAT AREN'T

Strange phantom animals (sometimes of a shape-shifting nature, and often not quite corresponding to usual zoomorphic norms) are alleged to have been seen during both fairy and poltergeist-hauntings, even into the present day. Indeed, sometimes old barguests were specifically spoken of as performing poltergeist or fairy-like pranks themselves. There was one notorious bogey from Northumberland known as the Hedley Kow, for example – it appeared sometimes in the form of a cow or foal and sometimes in the shape of a man or a truss of straw – which was alleged in local lore to trick servant girls outside by imitating the voices of their boyfriends before knocking over pails and pots, or else undoing their knitting and ruining their yarn inside the house. Nowadays the good people of Northumberland would surely blame a poltergeist for pulling such puckish pranks, however, and not the Hedley Kow.

The Hedley Kow was just folklore, but phantom cats and dogs have also popped up during many more genuine-sounding poltergeist cases down the years, some of which were blamed at the time upon fairies, and some of which were not. For example, inside the polt-haunted house of a man named George Lee in the Oxfordshire village of North Aston in 1591/92, amongst more typical ghostly phenomena, a series of "grotesque dogs" and other animals were observed. One time a "black object like a dog" appeared in the courtyard, and another time "a creature like a great brindled [streaked] dog without legs" was found in a tub used to sift meal in. The repeated use of the phrase "*like* a dog" here is surely significant. Just as interesting was the 1722 poltergeist haunting in Sandfeldt, Germany, where a race of "little crooked people" were supposedly busy teleporting innocent kids underground and offering them fairy-money to stay. Similarly fairy-like, however, was the vision seen by the children on 26 February of that year when they allegedly witnessed "a cavalcade of strange rough things almost like calves, but smaller" flying out of a shed, being flogged forward by a big man with a whip – bogey-beasts, in other words.

If one way in which bogey-beasts have been reconceptualised in modern times is through the prism of cryptozoology, however, then this is not the entire picture. Take, for example, the following description of a mystery beast; it was "a big animal, very heavily-muscled, with short legs" and "a bushy tail" which looked rather like "an exaggerated fox tail". Despite this fact, it had "a hyena's body", "a reddish colouring" and "short, stubby legs like a boar". It was seen hunting horses on a Utah ranch during April 1999 and, when pursued by the steeds' owner, simply "vanished into thin air". A matter for cryptozoology? You would have thought so. However, the above cryptid was seen during the notorious haunting at Skinwalker Ranch in Utah during the late 1990s and early 2000s, when, as is well known amongst ghost-aficionados, much poltergeist (and UFO) activity was also going on around the place. In such an outbreak of miscellaneous weirdness – which seems to have been a haunting of the land and sky in the area as much as it was of a mere farmhouse – the three contemporary disguised strands of fairy-lore would appear to combine to creative a narrative of what is often known nowadays as 'high-strangeness' or so-called 'window-areas' (zones in which a plethora of paranormal weirdness are alleged to occur with bewildering frequency).

It seems to me, then, that many thousands of people in the modern world still believe in fairies – it's just that, nowadays, they prefer to disguise this fact by using the word 'paranormal' instead. This probably shouldn't surprise us, seeing as the very word 'fairy' (or early versions of it, anyway) was originally an adjective used to refer to anything which seemed supernatural in any way shape or form … ghosts, impossible animals, or strange lights in the sky, for instance. Admit it, cryptozoologists, ufologists and ghost-hunters – some of those things you spend your time seeking are really just the fairy-folk in disguise!

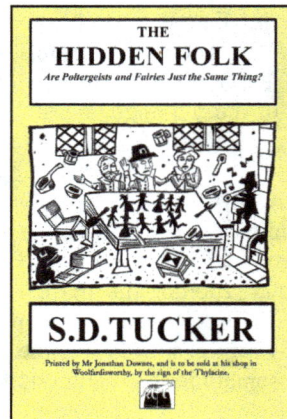

The Hidden Folk: Are Poltergeists and Fairies Just the Same Thing? by **SD Tucker** is available now from CFZ Press (ISBN: 978-1-905723-40-9)

THE MYSTERY OF THE MOORLAND WALLABY

Liam Dorricott

The red-necked wallaby is a medium-sized marsupial belonging to the Kangaroo family. Although native to the sub-tropical climate of Eastern Australia, variations of this species [Bennett's wallaby] can be found in the cool temperate climate of Tasmania. However over the decades there have been many reports of these animals in central England, rising to the matter of just how did they get so far away from the land down under?

In the western parts of the Peak District between the towns of Buxton and Leek, eyewitness accounts of a lone wallaby to an entire colony have been reported in the moorlands for over 70 years. From sightings on woodland edges to the rocky outcrop of The Roaches, as well as near-collisions with vehicles on the dark back roads, the tales of "fury-things" bouncing through the brush are not unheard of between the people who visit the area; and more surprising to the many people who remain sceptical, is that the eyewitnesses are not wrong.

Lieutenant Colonel Henry Brocklehurst, who served as a pilot during World War I,

released his captive red-necked wallabies into the wild after wartime [World War II] regulations forced the closure of private zoos in 1939.

The five wallabies naturalised quickly due to high vegetation and low threat from predators, and gradually the small group grew and separated into mobs and by 1960 there were an estimated 50 wallabies roaming this area of the Peak District, with a higher concentration in the western region near the Staffordshire border. But when the destructive winter of 1962/63 hit the region, a species well equipped for heat wave weather faced a drastic change in what it meant to survive, and the colony was feared to have been wiped clean from the moorlands.

But the wallabies lived on as sightings continued and in 1965 a young lecturer in zoology by the name of Derek Yalden became the wallabies' primary chronicler. He maintained an annual survey for many years and personally lead many expeditions to monitor the remaining wallaby population. But unfortunately his findings only confirmed his suspicions, that the wallabies were in decline.

Surveys conducted throughout the 1970's and 80's accounted only a dozen wallabies left in the wild and when another disastrous winter hit the district in early 1982 the population decreased again. But the weather wasn't the only cause in the population loss, it is thought that increased tourism in the area restricted the colony to grow.

By the mid-1990s it was believed the population was down to a number of three; however early into the new millennium [2003] an older female was recorded and another, younger female was recorded as late as 2009. There age was not documented but the recorded life span of a wallaby in the wild is around 10 years, meaning that these individuals could have been born either side of 2000.

Although there has been a steady decline of sightings since the last confirmed report in 2009, there are still reports of what may be wallabies in the Peak District to this day, with the last documented sighting in April 2015

(full report here: http://www.roaches.org.uk/wallabies.htm).

Until conclusive proof is brought forward, however, it must be presumed that the Peak District Wallaby followed in the footsteps of so many animal species across the world, and became extinct.

TASMANIA 2016 EXPEDITION: PRELIMINARY REPORT

Richard Freeman arrived back in the UK at the end of March. Two days later he came up to North Devon to return to stock the equipment that he had taken. Because we had issues of *Animals & Men* and the newsletter imminent, we decided to get a brief account from him for each of the publications. However, Richard has asked me to stress that his full account of the expedition will be written in time for issue 57 of *Animals & Men* and that he would hope that everybody realises that these accounts are just stopgaps!

This year's expedition was the smallest in terms of manpower, consisting only of CFZ

RICHARD FREEMAN

Bush fire damage in the northwest of Tasmania

Australia's Mike Williams, and myself. We found that smaller numbers make for an easier expedition because there are fewer disturbances when searching for animals, and it is less daunting for interviewees. Once again, Toyota had generously sponsored us by providing a car and a fuel card.

First we revisited the areas in the north-west where we had been in 2014. However, these had been badly burnt by bushfires; apparently there had been 28 lightning strikes in a single day, and the area looked as if it had had a visitation from Smaug. So we moved our focus to the central highlands.

We put a higher emphasis on night-driving and spot-lighting as most sightings of the Tasmanian wolf happen from cars at night, as the animal is crossing the road. A story about the expedition in a small local newspaper had brought forward a witness. As before, we are not revealing exact locations and witness names if they have requested anonymity. One of the best witnesses we talked to was an elderly man in a small town who had seen the creature himself in the 1950s, whose wife had seen it in the 1980s, and whose son had seen it just one year ago. In 1958 the man had been putting away his car, when in the half-light, what looked like a large dog walked down the road towards him. Thinking it was his

The tallest tree in Tasmania—a swamp gum nearly 300ft tall

Thylacine prints c.1938 at Hobart Museum

Thylacine skulls at Hobart Museum

neighbour's dog he tried to shoo the animal away, but it ignored him and walked right past him. Thinking it was strange, he switched on the car's lights, illuminating the striped hind-quarters and stiff tail of a thylacine. Previously, a man crop-dusting had reported seeing one, and soon after his neighbour saw the animal as well.

The man's wife saw a thylacine in the early '80s, approximately two miles from their home as it crossed the road in front of her. It seemed unconcerned about the car, and she clearly saw the dog-like head, stiff tail and striped hind-quarters. The man's son saw another around

Easter 2015. He was with his father on a forestry road. His father had gone into the forest looking for the remains of an old convict road. He heard his son shout that there was a strange animal on the road, the like of which he had nerve seen before.

He later described it as being the size of a whippet, with a long stiff tail, and a striped back. It emerged from the forest and walked up the road, no more than 20 feet away from him. It re-entered the forest, emerged again, and then once more went back into the forest. It seems to have then doubled back and circled round the father's car. It left a series of tracks,

which were photographed the next day. Having looked at the photographs, they do seem to resemble those of a young thylacine.

Mike and I camped out at this very spot for a considerable time. We baited camera traps with road-killed wallaby, but only succeeded in photographing quolls and Tasmanian devils. Night-drives and spot-lighting revealed many prey species such as wombats and wallabies.

The aforementioned witness was a retired logger, and recalls, in the 1950s, seeing what he thought were the footprints of large dog on a regular basis. He asked the foreman if anybody owned a dog, and the foreman that they were the footprints of a Tasmanian tiger. Another one of the workmen laughed at the

idea, but a few days later walked around a boulder and found one of the animals sitting there. It threatened him with a gaping yawn, and he beat a hasty retreat. The most recent sighting he knew of was just six months ago, when a group of men out spot-lighting saw a thylacine beside the road.

Another witness from the same town worked for a hydro-electric company. In the mid-1970s he and three other witnesses watched a thylacine from a car as it crossed the road and paused to look at them. They had the animal in view for about six minutes. Again it seemed unconcerned about the car. The man's cousin and several friends had a remarkable sighting just 18 months ago, when a male, female and three pups crossed the road in front of them.

DISCUSSION DOCUMENT: DUTIES FOR REGIONAL REPRESENTATIVES

The CFZ had had regional representatives for over twenty years now. Some of them have done remarkable things, some nothing at all, and some something in between. I originally intended my first wife to manage the list of regional reps, but as history shows, that never happened.

Ever since Alison and I split up I have been intending to ask someone else to take over the job, and finally a few months ago I got around to it. Ronan Coghlan has agreed to take over the onerous task, and has come up with a list of suggested roles for regional representatives, which I post here for public discussion.

[1] In the event of a reported sighting of a mystery animal in the representative's area, all possible data should be gathered and forwarded to CFZ. Likewise, news of further developments should be sent on as they occur.

[2] Representatives should try to discover if there were any sightings or other anomalous events in their areas in the past, but should only send on stories of UFOs or ghosts if they consider them important, as otherwise their is the danger of CFZ being swamped.

[3] Representatives should, if possible, look into local folklore to discover if stories of anomalous events in the area occur. Liaisons should be initiated with the Bird, Butterfly and Conservation Officer in their areas where possible. They should, in addition, try to gather an archive of Fortean zoological material from their local studies libraries.

[4] Representatives should initiate liaisons with groups dealing with anomalies and nature in the area, provided they consider them and their personnel suitable.

[5] Representatives should have the option of offering sales of books to local bookshops. However, some might find this distasteful and so this should not be regarded as an actual representative's duty.

Letters

The editor and his compadres welcome letters for publication on all subjects covered by this magazine. However, we would like to stress that neither this magazine, or the CFZ are responsible for opinions expressed, which are purely those of the letter writer.

A MINNESOTA MYSTERY SNAKE AND CHINESE FELINE CRYPTIDS

I have recently come across two stories of a cryptozoological nature which I thought I'd share with *Animals and Men* readers, although the first one was passed on to me by Minnesota based Facebook friend Greg Brick, who happens to be an expert caver and geologist.

I have never come across these stories anywhere else in any cryptozoological or Fortean literature. The first story concerns a large snake or crocodilian (in my opinion) called the "Buffalo" or "Buffeloe

Snake" ; I noticed both spellings in the piece of writing Greg shared with me in February 2016.

The Manuscript Journals of Jonathan Carver, 1760s, mention a story told by the Naudowessee Indians of a "monsterous" snake which they called the "Tautongo Omlisho" , which signified in English the Buffeloe Snake (sic) which had horns and four feet with claws like a bear. It was 3 fathoms long and as big and round as a buffalo. It had a black head and tail. The middle from the head to the tail was red with a fin running down the back. "The chief of the warriors of the Mottobauntoway band who was my particularly friend told me he saw one of these serpents on the plains with a young one which was in the crotch of a large tree by which he supposed the old ones climb up on the trees. The figure of this serpent I

have annexed to this journal." Apparently the "snake" also appeared in Carver`s Cave in St Paul`s ,Minnesota, according to Greg.

"In 1867 an investigator of Carver`s Cave reported finding an Indian pteroglyph of a snake there; he suggested that this was the "autograph" of Ottahtongoomlishcah." Ottahtongoomlishcah was the name of the chief. Here I have included my copy of the figure that appears on page 99 of Carver`s Journal, the edition Greg shared with me.

Now in the *South China Morning Post*

(Hong Kong) for October 11th 1934 "Strange Creatures Raid Village Tientsin, Oct 2nd 1934". Villagers of Hsiao Ma Chuen, a small village on the western borders of Chihli and about 40 miles south-west of Shihchiachwang, had a singularly disagreeable experience yesterday, when a horde of over 20 strange and nondescript beasts as big as calves and with feline heads and greyish yellow hides, came out of the neighbouring woods and descended upon the village, devouring practically all the cats, dogs, and fowls in the village and its immediate neighbourhood.

Fortunately the village children were staying in their homes and no one was killed. The attack was stated to have come so suddenly and unexpectedly that when the villagers succeeded in organising themselves into a defence unit with rods, sticks, and some with rifles, the unwelcome visitors had already beaten a quiet retreat into the woods.

Best Wishes
Richard Muirhead

A MANIFESTATION OF MONSTERS

EXAMINING THE (UN)USUAL SUSPECTS
DR. KARL P.N. SHUKER
FOREWORD BY KEN GERHARD

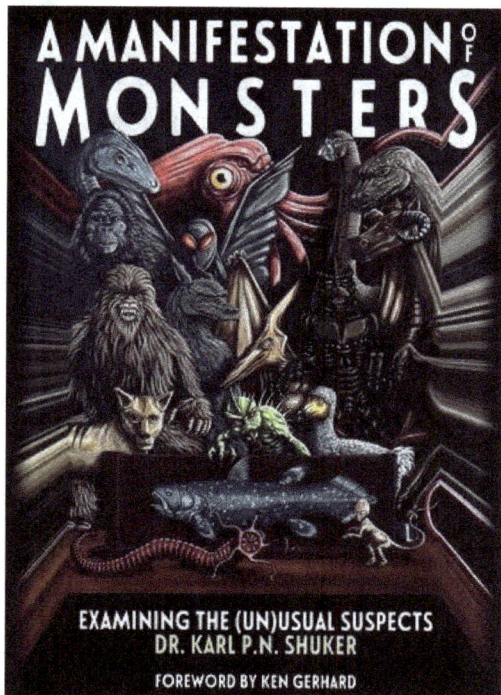

Paperback: 216 pages
Publisher: Anomalist Books (4 Sept. 2015)
Language: English
ISBN-10: 1938398521
ISBN-13: 978-1938398520

Dr Karl Shuker is surely one of the most influential writers in the fields of cryptozoology and forteana in modern times. His new book *A Manifestation of Monsters* does not disappoint and has a similar appeal to the writings of the late, great Ivan T Sanderson. Indeed, this latest tome puts me in mind of Sanderson's 'Things' and 'More Things'.

The book had its genesis in a spectacular painting by Michael J. Smith called 'Cryptids'. The picture shows a collection of unknown animals such as the Tasmanian wolf, the yeti, giant squid, moa, Mongolian death-worm and many others. It mixed probable flesh and blood cryptids with stranger creatures such as Moth-man and the chupacabras. Karl decided to write a book featuring these creatures, and a good deal more to boot.

As always Karl comes up with fresh material that will delight and intrigue readers. In the first chapter he takes a look at strange creatures held in early zoological collections. The first gorilla in the UK was held in a travelling menagerie where it was mislabelled as a chimp. I have always thought that the 'Girt Dog' of Ennerdale, that terrorised the Lake District in the early 19th century, was a Tasmanian wolf that had escaped from one of these horse-drawn zoos. Creatures labelled 'zebra wolves' were mentioned in the lists of creatures held. Yet here Karl has a suggestion that I had never before considered. Another strange name appearing in these lists is 'prairie fiends'. Karl, looking at the descriptions, concluded that some of these beasts were late surviving ground sloths, the 'prairies' being the pampas of South America.

As someone highly interested in dragon legends and lore, the chapter on giant lizards, unknown to me was the monster lizard known as the 'Afa' from the marshes of Iraq (now probably vanished along with its marsh home). Equally new was the jahoor, said to inhabit the swamps of India and to live in symbiosis with the saltwater crocodile. The m'o of Hawaii (where there are no indigenous land reptiles) was a thirty-foot, black scaled, dragon-like beast. It was allegedly seen by crowds of people in the 19th century.

Elsewhere we meet fire-spitting phantom Christmas pigs and pterosaur-like death dealing birds from Scandinavia, monster eels in Guam, giant fossilized burrows of prehistoric worms, giant frogs and a Bilbo Baggins link to the deathworm.

I won't reveal everything that is in the book as that would spoil it. But as with all of Karl's work *A Manifestation of Monsters* is a wild fairground funhouse of a book that will leave even those who have toiled long years in the groves oflong years in the groves of forteana scratching their heads.

Paperback: 368 pages
Publisher: Granta (8 Oct. 2015)
Language: English
ISBN-10: 1847088996

I love books. They have been the greatest indulgence in my life since I was about six years old, and although I subsequently discovered sex, alcohol, rock and roll, politics, narcotics and other indulgences, truthfully none of them have ever eclipsed books in my life.

Over the years my favourite writers have come and gone, but surprisingly not as dramatically as have my favourite musicians or my favourite drinks. My favourite authors have included C S Lewis and Gerald Durrell since the mid 1960s, but I am happy to say that I continue to C J Stone and John Higgs, for example, have made appearances in my writings over the years, and both have appeared at Weird Weekends, as I rant happily about their extraordinary prosecraft, and there have been others. But today I wish to introduce you to another of my favourite contemporary writers - Patrick Barkham.

But today I wish to introduce you to another of my favourite contemporary writers - Patrick Barkham.

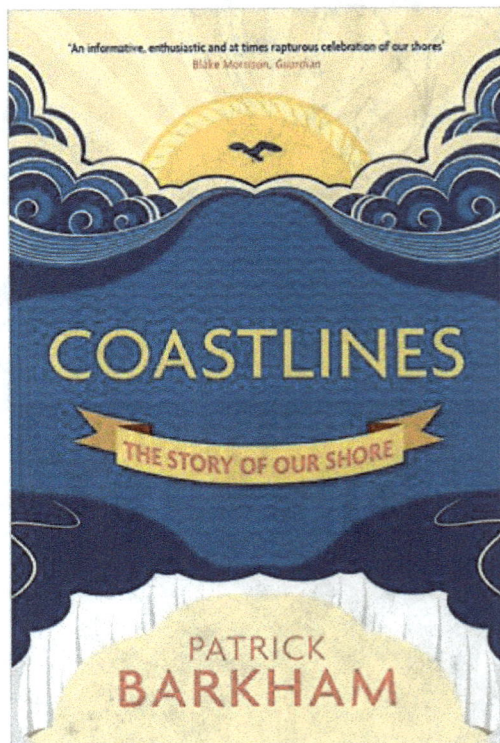

'An informative, enthusiastic and at times rapturous celebration of our shores'
Blake Morrison, Guardian

COASTLINES

THE STORY OF OUR SHORE

PATRICK **BARKHAM**

I first came across Barkham a few years ago when he wrote a book about his personal quest to see all the native British butterfly species in a single year. I have never made any secret of my fondness for the genre of lepidoptera memoirs written by people like L Hugh Newman and P B M Allan, but these are roundly things of the first half of the 20th century. Patrick Barkham brought the genre up to date with mentions of things such as irritating girlfriends, email and text messages, as well as the increasing levels of bureaucracy which surrounds the pursuit of the natural sciences as we progress further into the 21st century. But he did so whilst never losing touch of the sheer childlike wonder which is what drew people like him and me to the study of such things in the first place.

A couple of years back, when I was still producing my monthly webTV show, which I have every intention of resuming at some point before the universe and I are very much older, I interviewed

Barkham about the book and about the then concurrent spate of strange British butterfly sightings. A year or so later he published a book about badgers, and then - last year, I believe - came his third book. This time about the British coastline.

I have been deep in a love affair with the British coastline for half a century this year, ever since my family spent a month in Guernsey in 1966, and I discovered the awesome range of flora and fauna to be found in rockpools. So when I saw that Barkham had written a book on the subject of Britain's coasts (and no, I do not need to be reminded that The Channel Islands are zoologically and geographically damn all to do with the British Isles) I put it on my Christmas list, and my darling wife stepped up to the mark.

A brief anabasis here whilst I reproduce the blurb provided by the publisher.

"Told through a series of walks by the sea, nature columnist and author Patrick Barkham, *Guardian* columnist and acclaimed nature writer, explores Britain's beaches, coasts and cliffs in his latest book *Coastlines: The Story of Our Shore*. From smuggler's coves to Brownsea Island, witness the profound story of our island nation and how we are shaped by our shores."

Now, I must have been stoned when I first read about the book, because somehow I had got it into my head that this was going to be something like a grownup's version of the Ladybird book of the seashore that I had when I was on the beaches of Guernsey half a century ago.

But in fact it was something far more complex and many layered, and whilst I missed out on my planned Boxing Day of sitting drunkenly drooling over a dozen different pictures of different species of spidercrab, this book is so much more.

It is nothing short of a chronicle of all sorts of multifaceted aspects of the shores of these sceptre'd isles, covering economics, history, geography, politics, sociology, art history and folklore as well as the natural history that I had been expecting. Any book that covers the political backlash against

Thatcher, the social politics of the post-'Troubles' Northern Ireland, the history of French and Dutch incursions into the England of the seventeenth and eighteenth centuries, and even the more obscure knock on effects of the belief that a UFO crashed into Rendlesham Forest in 1980 truly has to be admired.

But you know the weirdest thing about this book? Probably ninety percent of this was stuff that I had not read before. But there was so much more that I would have put into it had I been the author. Hartland Point, Barmouth Bay and Westward Ho!, for example, as well as Saunton Sands, Dawlish Warren, Maenporth Beach and that bit just to the west of Teignmouth whose name I cannot remember, could all have equally well have fitted seamlessly into the narrative of this book. As could Canvey Island, the beach near South Shields where Blaster Jack built his home, and all sorts of places that I have come across during my multifarious travels.

None of this is a criticism of this book. Far from it. It is testament that Barkham took on an impossibly complex task and got away with it. It is just that these peculiar islands where I have lived since I was eleven, and where Barkham has lived since I was far younger than that, are such an impossibly complex place that nobody could possibly hope to recount all of the stories that are thrown up by them in one, or even a dozen, books if this size.

We are an island race, and these islands have enough stories to fill a library. Indeed they already have, many times over. But as a primer, this book is without equal.

However, if I may revisit the emotion behind the front cover of an issue of a Fanzine called 'Sniffing Glue' back in the day, this is a book to show you the way. Now go and find your own stories.

See you on the beach.

PS I would ALSO still like that book with a dozen different pictures of spidercrabs if anyone comes across something of the sort.....

www.ingramcontent.com/pod-product-compliance
Lightning Source LLC
Chambersburg PA
CBHW050602280326
41933CB00011B/1952